Books by Dr. Michael W. Fox

The Boundless Circle (1996) Quest Books, Wheaton, Illinois

Troubled Harvest: Seeds of Hope (1996) Good Earth Publications, Columbus, North Carolina

Superpigs and Wondercorn: The Brave New World of Biotechnology and Where It All Might Lead (1992) Lyons & Burford, New York

The Soul of the Wolf (reprint edition 1992) Lyons & Burford, New York

Understanding Your Dog (revised edition 1992) St. Martin's Press, New York

Understanding Your Cat (revised edition 1992) St. Martin's Press, New York

You Can Save the Animals (1991) St. Martin's Press, New York

Supercat (1991) Howell Books, New York

Superdog (1990) Howell Books, New York

Inhumane Society: The American Way of Exploiting Animals (1990) St. Martin's Press, New York

St. Francis of Assisi, Animals, and Nature (1989) Center for Respect of Life and Environment, Washington, D.C.

The New Eden (1989) Lotus Press, Santa Fe, New Mexico

The New Animal Doctor's Answer Book (1989) Newmarket Press, New York

Agricide: The Hidden Crisis that Affects Us All (1988) Schocken Books, New York

The Whistling Hunters (1984) State University of New York Press

Behavior of Wolves, Dogs, and Related Canids (reprint edition 1984) R. E. Krieger Publishing Co., Malabar, Florida

One Earth, One Mind (reprint edition 1984) R. E. Krieger Publishing Co., Malabar, Florida

Farm Animals: Husbandry, Behavior and Veterinary Practice (1983) University Press, Baltimore, Maryland

The Healing Touch (1983) Newmarket Press, New York (paperback version of *Dr. Michael Fox's Massage Program for Cats and Dogs*, 1981, Newmarket Press, New York)

Love Is a Happy Cat (1982) Newmarket Press, New York

How to Be Your Pet's Best Friend (1981) Coward, McCann & Geoghegan, New York

Returning to Eden: Animal Rights and Human Responsibility (1980) Viking Press, New York

Understanding Your Pet (1978) Coward, McCann and Geoghegan, New York

Between Animal and Man: The Key to the Kingdom (1976) Coward, McCann and Geoghegan, New York

Concepts in Ethology, Animal and Human Behavior (1974) University of Minnesota Press

Integrative Development of Brain and Behavior in the Dog (1971) University of Chicago Press

Editor of:

The Wild Canids (reprint edition 1983) R. E. Krieger Publishing Co., Malabar, Florida

On the Fifth Day: Animal Rights and Human Ethics (1977) Acropolis Press, Washington, D.C. (with R. K. Morris)

The Dog: Its Domestication and Behavior (1977) Garland Press, New York

Readings in Ethology and Comparative Psychology (1973) Brooks/Cole, California

Abnormal Behavior in Animals (1968) Saunders, Philadelphia

Advances in Animal Welfare Science (annual series 1984–1987) Martinus Nijhoff in Holland, and The Humane Society of the United States, Washington, D.C.

Children's Books:

Animals Have Rights Too (1991) Crossroads/Continuum, New York

The Way of the Dolphin (1981) Acropolis Books, Washington, D.C.

The Touchlings (1981) Acropolis Books, Washington, D.C.

Lessons from Nature: Fox's Fables (1980) Acropolis, Washington, D.C.

Whitepaws: A Coyote-dog (1979) Coward, McCann & Geoghegan, New York

Wild Dogs Three (1977) McCann & Geoghegan, New York

What Is Your Dog Saying? (1977) M. W. Fox and Wende Devlin Gates, Coward, McCann & Geoghegan, New York

What Is Your Cat Saying? (1977) M. W. Fox and Wende Devlin Gates, Coward, McCann & Geoghegan, New York

*Ramu and Chennai** (1975) Coward, McCann & Geoghegan, New York

*Sundance Coyote*** (1974) Coward, McCann & Geoghegan, New York

*The Wolf**** (1973) Coward, McCann & Geoghegan, New York

Vixie, the Story of a Little Fox (1973) Coward, McCann & Geoghegan, New York

*Best Science Book Award, National Teachers' Association

**Nominee for Mark Twain Awards

***Christopher Award for children's literature

The Hidden Farm and Food Crisis That Affects Us All

Second Edition

Dr. Michael W. Fox

KRIEGER PUBLISHING COMPANY
MALABAR, FLORIDA
1996

Original Edition 1986
Second Edition 1996

Printed and Published by
KRIEGER PUBLISHING COMPANY
KRIEGER DRIVE
MALABAR, FLORIDA 32950

Library of Congress Cataloging-In-Publication Data
Fox, Michael W., 1937–
Agricide : the hidden farm and food crisis that affects us all / Michael W. Fox. —2nd ed.
p. cm.
Includes bibliographical references (p.) and index.
ISBN 0-89464-945-0
1. Livestock—United States. 2. Animal industry—United States. 3. Livestock factories—United States. 4. Animal welfare—United States. 5. Agriculture—United States. 6. Agricultural industries—United States. 7. Agricultural ecology—United States. 8. Agricultural pollution—United States. 9. Food industry and trade—United States. 10. Diet—United States. I. Title.
SF51.F69 1996
338.1'6'0973—dc20 95-2338
CIP

10 9 8 7 6 5 4 3 2

Contents

Preface to the Second Edition / xi
Introduction / xiii
1 The Farming Business / 1
2 The Economic Dimension / 26
3 The Health of the Earth and Its People / 50
4 The Nutritious Diet / 99
5 The Matter of Conscience / 108
6 Measuring the Results / 136
7 Toward a Saner Future / 152
8 Breaking the Cycle of Poverty and Famine: The Role of Humane Sustainable Agriculture / 163
9 Genetic Engineering and Our Farming Future / 185
Epilogue: Strategies for Change / 196
Notes / 219
Bibliography / 229
Supportive and Supplemental References / 235
Index / 249

for Mara

We don't inherit the land from our ancestors,
we borrow it from our children.

—Pennsylvania Dutch Saying

Preface to the Second Edition

The validity of the facts and figures compiled in *Agricide* has not changed since this book was first published. They stand as testimony and evidence of an increasingly dysfunctional agribusiness food industry that has forced thousands of family farms off the land and devastated the social economy of rural America. This industry has gained a monopolistic hold over how and what farmers farm, causing extensive ecological damage as a result of chemically dependent monocrop farming and intensive livestock and poultry production, along with animal stress and suffering.

While the introduction and Chapters 1–7 remain intact, I have added supportive and supplemental references to the bibliography in this second edition to provide additional resources and documentation. Also two new chapters have been added.

In Chapter 8 the negative consequences of *agricide* are looked at from a more global perspective, and the necessary steps to help alleviate world poverty and hunger by adopting the principles and practice of humane sustainable agriculture are detailed.

Chapter 9 discusses the ramifications of a new frontier in agriculture, namely, genetic engineering biotechnology.

As will be shown, if this new technology is not applied in accordance with the ethics and philosophy of organic and sustainable agriculture, then it will do more harm than good.

In spite of changes in administration and agriculture programs and policies over the past decade, it is evident that the U.S. government continues to be the servant of agribusiness rather than of the people. Consumers must therefore not look to government as the guarantor and sole arbiter of food quality and safety. Rather, we must all accept the reality of agricide and assert our power as consumers to transform industrial agriculture into a food production system that is healthful, ecologically sound and sustainable, equitable, and humane. How we the people can accomplish this revolution is outlined in a new epilogue that effectively transforms *Agricide* into a handbook for social change.

Introduction

EFFICIENT, regenerative agricultural productivity ensures the independence and security of any nation. However, the industrialization of American agriculture has recently emerged as a threat to this nation's security. This is not simply because agriculture is used as a major item of international trade ("agripower") in return for imported nonessential consumables and for the raw materials and income needed for the manufacture of domestic nonessential consumer products, inappropriate technologies, and military weaponry. It is also economically unsound in the long term because it is ecologically unsound. Agribusiness's industrialized exploitation of the land for commerce is in part responsible for widespread soil erosion, the depletion of deep-water aquifers, the deterioration of soil quality, and the pollution of our water and food. And in the process we have become increasingly dependent upon imports of fertilizers (such as potash and phosphates) and upon natural gas (for the manufacture of nitrogenous fertilizers). Nonrenewable resources are being wasted: agribusiness is "stealing" from the future.

The oscillations in world production and demand for agricultural produce create an unpredictable market, which causes great problems and anxiety for American farmers. The government assistance required to rectify the annual fluctuations in production and export demand creates a significant financial burden on the public. The only beneficiary is agribusiness and its allied multinational corporate interests. The general rubric "agribusiness" includes the petrochemical industry (which manufactures pesticides and fertilizers, and which now monopolizes much of the world's agricultural seed stock); equipment manufacturers; the grain, livestock-feed, and pet-food industries; the pharmaceutical industry; the wholesale food industry; and large corporate farms and contract producers.

A secondary beneficiary of current agribusiness practices is the medical industry, which profits from treating health problems that may be related to agrichemical and animal-drug residues in food, as well as the unhealthful dietary habits of consumers (encouraged in part by some food manufacturers).

Agricultural productivity is being boosted not only at the expense of resources, but at the expense of environmental quality and consumer health, as consequences of the overapplication of pesticides and drugs that are given to farm animals, many of which are now recognized as introducing serious consumer-health hazards. Efficiency is the name of the game, to keep prices low and to maintain a competitive edge in the world market. There is little altruism involved here to "feed the hungry world."

The quest to increase efficiency is pursued so single-mindedly that agribusiness factory farming is actually less efficient per acre and per animal than more traditional and ecologically sound farming practices, but it is profitable *overall* because of the advantages of economies of scale and the hidden costs that are paid for by a wholly uninformed public.

Automation on the corporate factory farm (whose capital-intensive, rather than labor-intensive, farming is especially remunerative, given the present unequitable tax structure) has resulted in massive unemployment and the destruction of rural communities and traditions. The human work force is one of agribusiness's many abused "natural" resources.

Shareholders want quick returns on their investments, so agricultural practices are not based upon ecologically sound, long-term planning—this is another significant factor in the agricide scenario. The land is literally being "mined," sucked dry, and polluted. The long-term costs will be far greater than the short-term profits and benefits that agribusiness claims to be reaping for the good of all.

The public pays the costs of federal and state regulation, monitoring, research, and education. And in order to reduce the national debt, the Reagan administration ill-advisedly emasculated the effectiveness of the regulatory agencies, by reducing compliance standards and associated costs of agribusiness (on the naive assumption that agribusiness will act with enlightened self-interest), such that the crisis of American agriculture has become intensified. Capitalism, without some vision of what constitutes enlightened self-interest, ultimately cuts its own throat. The first to suffer are the consumers and the thousands of farmers who have gone into receivership in recent years.* And then we must also consider the suffering of farm animals, who are being intensively raised in overcrowded "factory" conditions.

Against this background, agribusiness's claim that American consumers pay less for their food than any other people in the world ($3.00 for a broiler chicken) is only a half-truth. The indirect costs to consumers, in both financial and health terms, and the costs to the environment and to intensively

*Four hundred thousand people left farming in 1985. According to the Census Bureau, farmers now constitute only 2.2 percent of the U.S. population.

raised farm animals are astronomical. If the structure and practice of agriculture is not quickly changed, and made ecologically more sound, health-inducing, humane, and equitable, the next generation will face inevitable agricide.

Neither the public nor the disemployed or anxiously employed farmers are the beneficiaries of the industrialization of agriculture. The corporate entities that now have almost complete control of our government are pushing us to the brink of a world war over competition for, and monopoly of, world resources and markets.

The ironic tragicomedy of this scenario is that there has been no deliberate conspiracy on the part of agribusiness. It believes that it is right—in every sense of the word—that the system works, since it is profitable, and that any and all problems can be rectified by still more research and technological innovation. This is the comedy. The tragedy is that agribusiness seems incapable of making the paradigm shift in awareness to see that agricide is inevitable and can only be averted through enlightened self-interest. I am hopeful that, given the evidence, a transformation is possible, since there are many reasonable persons and institutions involved in agribusiness, and the political pressure from an increasingly informed public and media is intensifying. The future of this great nation may yet be secure, once the importance of a regenerative agriculture, rather than an extractive and destructive "agri-industry," is realized.

This book is not to be construed as an attack on the family farmer, animal scientist, veterinarian, and others dedicated to the efficient production of wholesome food for domestic consumption and for export. Nor is it—as some critics of my position have stated—a Communist-inspired (and -funded) plot to destroy American agriculture. Rather, it is a critique to convince all sectors of agribusiness, as well as consumers, that a system of agriculture has evolved over the past twenty years which is so flawed as to be ultimately self-destructive.

The statistics used in the book are from a variety of sources published at different times.* I have tried to use the most current available. And it must be remembered that experts often disagree. What is important for our purposes is not the numbers but for the general reader to grasp the magnitude of the problems addressed here.

Culture and industry have recently become mutually exclusive, the latter being developed at the expense of the former. However, the future of both depends upon their reintegration through land reform and greater equity; personal and corporate responsibility; and respect for the sanctity and dignity not only of the land and the lives of farm animals, but of those family farms and rural communities that are the backbone of a sound and self-sustaining agriculture. Now they, along with the highest values and potentials of our culture, are on the verge of extinction.

The longer the restoration of culture and agriculture is delayed, the more difficult the task becomes. Agribusiness's denials, rationalizations, and vested financial interests are not insurmountable obstacles, since this nation is not yet devoid of men and women of courage and integrity who are concerned and committed to serving the greater good. The reintegration of culture and industry, of regenerative agriculture and agricultural commerce, is possible once the self-interest of industry becomes consonant with the interests of society.

*While these statistics deal primarily with U.S. agriculture, this book is relevant to all countries that have adopted similar agricultural practices and livestock and poultry production methods.

1

The Farming Business

PUT away your children's picture-book images of farm animals running free in a sunny farmyard or in green pastures beneath shady trees. You may see a few cows, steers, and sheep in the fields and out on the open range, but not all of them are kept this way. Billions of chickens never see the light of day anymore, nor do millions of pigs and veal calves. The fields are growing more and more silent, as long, prefabricated buildings proliferate from coast to coast. Lights are kept on day and night to speed productivity in some animals, while others are kept in near darkness to reduce stress from overcrowding. In these confined quarters, behind closed doors, are the creatures we will ultimately consume. These buildings are the factory units of modern livestock farming and the animals within are the machines, converting plant foods and industrial byproducts into flesh, eggs, and milk.

"Factory farming" is the label that has been given to most modern systems of livestock production. In many respects this is appropriate, since the modern farm does look like a factory, with its huge grain silos and long, often windowless metal sheds. On a first trip to such a farm, you might take it for a military camp or a concentration camp, except that there are no uniformed guards. The buildings may be surrounded by a high fence and locked gate with NO TRESPASSING signs all around. No, it's not a secret military installation. The high security is to keep away people and stray animals who might be carrying some disease that could infect the stock.

But where are the animals? You can't see or even hear them. Ventilation fans in the buildings that go on and off automatically may give you a whiff of something organic, but it's not a familiar farm smell. That is because the animals crowded in the buildings have no bedding, so there's no rich-smelling manure made up of straw, urine, and dung, as in the old days.

You might see a big lagoon or two some distance from the buildings. The water is dark, still, and lifeless, except for a little algal bloom around the edges. This is where the animals' urine and feces are flushed, to be gradually decomposed.

On some farms, this liquid waste is recycled and given to the animals to drink, and the ooze at the bottom of the lagoons is sucked out from time to time and sprayed on the fields where crops are grown to feed the animals in the long metal buildings. Once they used to roam free in those fields, but now more and more pastures are plowed up to grow animal feed. In this way, more animals can be kept on the farm. Few farms can provide all the feed that is needed, so much is shipped in from out of state or comes ready-mixed from the feed company.

You might be lucky and visit a factory farm with one or more of the latest innovations. A centralized computer monitors each animal "unit" and is programmed to mix the right feed for each group. At scheduled intervals, feed is piped to the animals, and a machine automatically scrapes the floor clear of excrement. You might see a washing and filtering system that circulates the urine and water or feces back to the animals to drink and eat, or the equipment may wash and dry the manure for the farmer to mix into the animals' rations later. The motive behind this setup is efficiency: not all of what the animal eats is digested the first time around.

Manure used to be used on the fields, but now most farmers prefer to use costly but labor-saving chemicals, and not all use the slurry from the lagoons. Some put the excrement from the long sheds into deep, sealed pits to make methane gas, providing energy to heat the buildings and run some of the equipment. The bacterial sludge from these pits may be fed back to the animals.

Few visitors will be allowed into the buildings where the animals are kept. Ostensibly, this is because you might be carrying some disease which the animals, being confined and overcrowded, could easily contract in such close quarters. Some confinement houses, as they are called, have a glass viewing port for visitors and for the farmer to eyeball his animals once a day, because there's no need to go in with the animals: automation provides them with food, water, and occasional changes of air.

On some farms, particularly hog farms, you may be asked to shower and change into special clothes before you go into the building itself.

When you enter, you may see very little because there are no windows and the artificial illumination is kept extremely low. After a while, when your eyes have adapted to the half-light, they may begin to water because of the dust and ammonia in the air. In the worst of these confinement houses, you may gag, or be unable to see if you wear glasses or contact lenses.

Then you scan the area for the animals. What you will see next depends upon the kind of factory you are visiting.

At an egg factory, you will see thousands of cages arranged in long avenues. Sometimes they are piled high up to the roof. There may be forty to seventy thousand birds in each building. Each bird has little more than a 6″ × 7″ floor space, four and even five birds being crammed into a 12″ × 18″ cage; they live in the cage from twelve to eighteen months. The farmer may tell you that they like to be together and that keeping them packed in helps keep the building warm. But if you look closely, you will see that the ends of the birds' beaks have been removed. This is to prevent the birds from pecking each other to death. And if you went there at morning cleanout time, you would see a cart piled high with dead and dying birds. Some have been pecked to death; others are simply "burned out" from the demands of production or have died from social stress and disease. The birds look healthy, with bright red combs and wattles, but that's because they are given arsenic (which is known to cause cancer in man) in their feed to improve their looks and stimulate their appetites.

In some egg factories, the birds are starved and denied water for up to ten days. This shock treatment, which kills many birds, is done to force them to molt and stop laying eggs. After a rest period, they are "built up" again, and a second egg-laying cycle begins. However, not all farms follow this practice. Instead, the birds are killed after their first laying cycle and made into soup and pet food.

On a chicken (or broiler) factory farm, you enter a long shed, perhaps 450 feet long, containing thousands of birds with less than one square foot of floor space per bird. (The unwanted male chicks have often been slowly suffocated to death in plastic bags.) Unlike the laying hens, these birds at least are free on the floor, which is called a "deep-litter" floor, composed of wood chips or corn husks and the birds' excrement. This soaked litter creates an irritating environment, especially in humid weather, leading to contact burns on the birds' skin, called breast blisters, and respiratory problems from the high levels of ammonia that build up. Catching these birds for transport to slaughter is no easy task and results in a high incidence of injuries, especially broken wings and legs. Because of these problems, research has been done on raising the birds in cages, but they develop abnormally, with deformed limbs and fragile bones that break easily when they are handled.

The animals panic as you enter and kick up such dust that this, combined with the humid, ammoniated air, makes you retch and run to the nearest door. The farmer is amused at your reaction, and he may tell you that you would get used to it after a while if you worked there. But, like many hog farmers, he may suffer respiratory diseases from working only a few hours a day in such an environment. And what of the animals that live all the time in such conditions? Without constant medication in their feed, they would quickly succumb to disease.

The long building you entered might instead contain pigs being fattened for slaughter. As you walk down the central aisle, a double row of pens extends as far as the eye can see. On each side of you, groups of ten to forty or more pigs are eating, drinking, squealing, or lying half asleep on a metal or concrete slatted floor. Each pig may have less than eight square feet of floor space in a pen containing twenty or thirty others, and all have had their tails cut off. This is to stop them from cannibalizing each other, you are told. The fans suddenly go on, temporarily alleviating the heaviness of the air. You leave with the odor of the pigs in your clothing and your memory impressed by so much life being contained so efficiently in so little space of concrete and iron.

A nearby building might contain two hundred breeding

sows in iron crates so narrow they cannot even turn around, or they may instead be tethered by chains to the concrete floor in front of their food hoppers. When they are ready to give birth, they are put into metal farrowing crates that similarly restrict almost all activities: they can only stand up, lie down with difficulty, eat, drink, and evacuate.

Such immobility and confinement results in many problems, including lameness, arthritis, obesity, infertility, and birth difficulties. Sows normally make a nest before they give birth, and frustration or privation of this and other basic instincts can cause stress and "neurotic" behavior patterns: excessive chewing or chomping on the bars of the crate or pen, for example.

Piglets are removed from the sow often as early as three weeks after they are born and may be placed in wire battery cages, just like laying hens, or on flat deck-pens sometimes stacked in three tiers and providing each piglet with less than two square feet of floor space. Such overcrowding can be extremely stressful and increase susceptibility to disease, hence the piglets are usually under a regimen of constant medication via their food. Overcrowding can lead to fighting and boredom vices such as tail chewing, and in order to reduce this vice, which may lead to cannibalism, the piglets' tails are cut off.

Of all the various animal factories, a veal factory might leave you with the most favorable impression. Here, the air has to be kept fresh because the calves are so susceptible to pneumonia. You see clean little stalls, each containing a calf, with its large eyes staring innocently out at you. One may reach out and suck your fingers as you stroke its muzzle. The place smells milky and warm with the bodies of the calves. As you leave, you notice (but it doesn't quite register with you) that the farmer has switched off all the lights. There are no windows, and the calves are left in total darkness except for two hours each day, when they are fed a milk-substitute liquid that is deficient in iron, in order to keep their flesh pale and anemic. These are sickly creatures, grown in total darkness for sixteen weeks. Some veal producers now provide calves with light and more iron in their liquid feed. But still they are denied any hay or straw, even though they crave such roughage, because it would darken their meat. Their feed is laced with antibiotics to control disease.

Outside, you breathe deeply and drive away, past more silent factories and fields devoid of animals. Not all American and European farms have become factories, and not all farmers deprive their animals to the degree I have described. Not all, but most. Virtually *all* layer hens, broiler chickens, and veal calves are now raised in these total-confinement buildings, as well as more than half of all pigs produced each year. Their numbers are difficult to visualize. Nearly 3.5 billion broiler chickens, 280 million battery hens, 1 million veal calves, and 50 million or more pigs are raised under the conditions I have depicted, which are sometimes better or much worse than I have detailed.

Beef cattle in the United States and other countries are finished (fattened) for 100–120 days in dirt lots, which are often overcrowded and lack any shelter from the elements. The cattle are fed a "hot" diet, high in concentrates and grains, which is stressful since cattle are basically consumers of roughage. This diet can lead to bloat, foot inflammation (laminitis), acid indigestion, and ulceration of the digestive system with subsequent infection of the liver. In order to reduce injuries in these overcrowded feedlots, the cattle have their horns cut off—without an anesthetic. They are also castrated without anesthetic and branded with a red-hot iron. These brutal treatments have adverse effects on productivity, but because of mass production—the economies of scale—there is little economic incentive to change such practices or to adopt more humane alternatives.

It is in order to keep food prices down in the face of rising land, labor, and production costs that farmers have adopted these and other inhumane and unethical practices. But tethering for an entire lifetime by a chain nailed to the ground is no acceptable way to treat any animal, no matter what economic justifications are given. Giving them more space and a more rich and varied environment is somewhat most costly, and sanitary conditions are harder to maintain. These problems represent a new challenge, and more research is needed to design systems that are sanitary, reduce the spread of disease, and are cost-effective and consonant with the animals' psychological needs. Such research has already been undertaken in Europe and Great Britain, and a handful of farm-animal scientists in America are at last beginning to study humane alternatives in livestock and poultry husbandry.

Jim Mason and Peter Singer vivify what the factory farming of animals entails:

> Regardless of the type of animal confined or the commodity purchased, all factory systems are designed to make more money from more animals. Instead of hired hands, the factory farmer employs pumps, fans, switches, slatted or wire floors, and automatic feeding and watering hardware. The factory farmer is a capital-intensive farmer whose greatest investment is in time- and labor-saving equipment. Success in farming is not achieved by direct care for the animals. It does not depend on the well-being of individual animals or even on individual productivity. Success comes from maximally efficient use of equipment. It is measured by year-end production records. Like managers of other factories, capital-intensive farmers are principally concerned with cost of input and volume of output. A certain amount of wastage doesn't

> matter if the product wasted is cheap by comparison with overheads and if eliminating the wastage would raise costs or reduce output. All this is as true of animal factories as of any other factory; the difference is that in animal factories the product is a living creature capable of pain and fear, a creature worthy of moral consideration that inanimate objects neither require nor could benefit from.[1]

While the humaneness of farm factories will be discussed later, there are a number of practices associated with such operations that require immediate consideration.

There is no question that confinement and overcrowding produce physical stress in the animals so treated. The stress sometimes leads to aggression, and so pigs' tails are docked and chickens are debeaked and sometimes declawed in the manner described.

The unnatural conditions in which the animals are kept raises medical problems.

> Intensive animal production creates housing problems in which much needs to be learned with regard to temperature, humidity, light, air circulation, space requirements, type of flooring, waste disposal and many other problems. As an example, swine under confinement show an increase in foot and leg problems; cracking of the hoof and stiffness occur frequently. Whether this is due to nutritional deficiencies, mechanical injury, infection or other causes is not definitely known.[2]

Mortality rates on many hog farms can be as high as 30 percent (and 80 to 90 percent of the pigs may have signs of pneumonia when inspected after slaughter). They average about 20 percent on dairy and veal-calf farms. Up to 30 percent of battery laying hens are culled during their short 1–1½ years in cages. Some 116,000 pigs, cattle, and sheep and 15 million poultry are condemned for human consumption prior to entering federally inspected slaughterhouses; 140,000 tons of poultry meat is condemned annually by inspectors at the slaughterhouses, along with 325,000 whole carcasses and 5.5 million major parts of hogs, sheep, cattle, and calves.[3] Liver abscesses alone, in feedlot cattle, mean a loss of $10 million annually of food that could be consumed.

Infectious diseases can quickly reach epidemic proportions in high-confinement density, so the animals are routinely fed a wide array of drugs. It is ironic that farmers can buy antibiotics and other drugs over the counter at their local feed and supply store and from traveling salesmen, and at a nominal price compared to what the consumer pays the pharmacist for the same drugs that are available only on prescription. (Even pet owners pay more and must have a prescription from their veterinarian.)

Many drugs are being sold over the counter by veterinarians or their employees in violation of laws requiring prescriptions for their purchase and without regard for their intended use. Apparently undercover agents were able to purchase prescription-only drugs in this manner in thirty-three of thirty-seven veterinary clinics visited in eastern Iowa without talking to a veterinarian or informing him of what the drug was going to be used for, according to one regional Food and Drug Administration (FDA) enforcement official.[4] Clearly because of the enormity and complexity of drug use in farm animals, the FDA is incapable of coping with the problem.

While confinement can reduce the high losses associated with parasitism, other diseases, particularly epidemic viral infections, increase costs and necessitate more rigorous preventive-medicine programs. New disease complexes have appeared, involving nutritional imbalances or deficiencies coupled with environmental stress (e.g., high ammonia levels), social stress, and resistant pathogenic organisms.

Since the aim of farm factories is to produce goods as quickly and cheaply as possible, the animals are fed abnormal diets to achieve the desired results. Some of America's vast surplus of milk and grain is sold cheaply to the veal and swine industries. Grains are fed to livestock and poultry to make their flesh more plump and so that they can be killed at an earlier age on this rich but often harmful diet. On such a diet, cattle, for example, develop acute indigestion, ulcers, liver abscesses, and a painful disease of the feet, and are also more susceptible to infections and metabolic diseases.

The use of antibiotics in animal feeds has been banned in the United Kingdom, West Germany, Holland, Sweden, Norway, and Denmark. Yet it continues in the United States. In testimony before a House Government Operations Subcommittee on July 24, 1985, it was revealed that the FDA has no comprehensive lists of an estimated 20,000 or more brands of animal drugs currently being used to boost farm-animal productivity and control disease. Only 2,500 have been approved by the agency. Committee investigators also found that some 3,500 of the unapproved drugs can leave residues in meat, milk, and eggs, and are capable of causing cancer or birth defects in humans who eat these foods.

According to one government study,[5] almost all poultry, ninety percent of all pigs and veal calves, and sixty percent of all cattle receive antibacterial additives* in their feed; and seventy percent of all beef cattle are given growth-promoting additives.† The General Accounting Office has identified over 140 drugs and pesticides that are likely to remain after slaughter as residues in meat.[6] In this study, we are mainly concerned with drugs used in animal feeds (referred to as

*Antibiotics, sulfas, nitrofurans, and arsenicals.
†Hormones, amino acids, vitamins, minerals.

"medicated" feeds) as growth promoters and to improve disease resistance.

Twenty-one drugs are approved by the FDA for use in animal feeds. Of these, seventeen are antibiotics and four are other additives.[7]

The seventeen feed-additive antibiotics are classified into systemic and nonsystemic antibiotics. The systemic antibiotics are of principal concern, because they are the ones that can be absorbed in significant amounts from the small intestines and can thus lead to appreciable residues in edible animal products. For growth promotion in newly born animals or newly hatched avians, levels of systemic antibiotics range from 2 to 50 grams per ton of feed (about 2.2 to 55 parts per million). For control of systemic disease for the newborn, they range from 200 to 1,000 g./ton (about 220 to 1,100 ppm).[8] Systemic antibiotics include the tetracyclines, erythromycin, lincomycin, and penicillin.

According to J. L. Krider, there were approximately sixty million tons of commercial feeds produced in the United States in 1969; about forty million tons, or two-thirds, of this feed contained additives.[9] The FDA estimates that 77.8 percent of the meat and eggs consumed in the United States are derived from animals that were fed medicated feeds.[10]

The amounts of antibiotics used in the United States for the purposes of animal growth promotion and feed efficiency have increased dramatically, according to Clark Burbee, from about 265,000 pounds in 1951 to 12.3 million pounds in 1978. Burbee also reports that almost half of the antibiotics produced in this country are now used for low-level supplementation to livestock and poultry feed, and that about half of this is penicillin and tetracycline;[11] John Elliott puts additions to animal feeds closer to 40 percent of American antibiotic use.[12]

In another set of comparisons, antibiotic production in 1960 was shown to be 4.16 million pounds; 2.95 million pounds (or 71 percent) were used for human and animal therapy.[13] The balance was added to animal feeds for growth promotion and disease prevention. By 1970, total production of antibiotics had increased fourfold, to 16.9 million pounds. Of this amount, 9.6 million pounds (57 percent) were for human and animal therapeutic purposes, and the remaining 7.3 million pounds were for use in animal feeds. According to the U.S. International Trade Commission, the five-year annual average production of antibiotics in the United States during the 1971–75 period was 18.94 million pounds.[14] Medicinal uses accounted for 11.16 million pounds, or 59 percent. The remaining 7.68 million pounds (41 percent) were used as feed additives and for other nonmedicinal purposes.[15]

Drugs used as growth promoters, disease inhibitors, and to improve feed efficiency are given to livestock in their feed or water, or by injection. Following is a list of the standard dose of drugs employed for poultry and the three major classes of livestock.

TABLE 1

CATTLE

Feed Efficiency	Use Level
Bacitracin zinc	35–70 mg./head/day
Chlortetracyline	70 mg./head/day
Erythromycin	37 mg./head/day
Melengestrol acetate	0.25–0.50 mg./head/day
Monensin	complete feed: 5–30 g./ton of feed
	feed continuously: each animal must receive not less than 50 or more than 360 mg. daily
Oxytetracycline	75 mg./head/day (finishing cattle)

Growth Promotion	Use Level
Bacitracin	35 mg./head/day
Bacitracin zinc	35–70 mg./head/day
Chlortetracycline	70 mg./head/day
Erythromycin	37 mg./head/day
Melengestrol acetate	0.25–0.50 mg./head/day
Oxytetracycline	75 mg./head/day (finishing cattle)

Iodine Source	Use Level
Ethylenediamine dithydriodide	400–500 mg./head/day for 2–3 weeks (not to be administered to dairy cattle in production)

Milk Production	Use Level
Oxytetracycline	75 mg./head/day
Thyroprotein	0.5–1.0 g./100 lb. body weight

CHICKENS

Bloom and Feathering	Use Level
Arsanilic acid or Sodium arsanilate	0.005–0.1% (45–90 g./ton)

Egg Hatchability	Use Level
Bacitracin	100 g./ton
Bacitracin methylene disalicylate	100 g./ton
Bacitran zinc	100 g./ton
Chlortetracycline	50–100 g./ton
Oxytetracycline	50–100 g./ton

Egg Production	Use Level
Arsanilic acid or Sodium arsanilate	0.005–0.01% (45–90 g./ton)
Bacitracin	50 g./ton first 4–6 weeks of egg production; 10 g./ton remainder of laying period
Bacitracin methylene disalicylate	50 g./ton first 4–6 weeks of egg production; 10–50 g./ton remainder of laying period
Bacitracin zinc	50 g./ton first 4–6 weeks of egg production; 10 g./ton remainder of laying period

Chlortetracycline	50–100 g./ton
Erythromycin	18.5 g./ton
Oxytetracycline	50–100 g./ton
Roxarsone	0.0025–0.005%

Eggshell Texture and Quality	Use Level
Oxytetracycline	50–100 g./ton

Feed Efficiency	Use Level
Arsanilic acid or Sodium arsanilate	0.005–0.01% (45–90 g./ton)
Bacitracin	4–50 g./ton
Bacitracin methylene disalicylate	4–50 g./ton
Bacitracin zinc	4–50 g./ton
Bambermycins	1–2 g./ton
Chlortetracycline	10–50 g./ton
Erythromycin	4.6–18.5 g./ton
Furazolidone	0.00083–0.0011% (7.5–10 g./ton)
Lincomycin	2–4 g./ton
Oleandomycin	1–2 g./ton of complete poultry ration
Oxytetracycline	5–7.5 g./ton
Penicillin	2.4–50 g./ton
Roxarsone	0.0025–0.005%
Tylosin	4–50 g./ton
Virginiamycin	5 g./ton

Growth Promotion	Use Level
Arsanilic acid or Sodium arsanilate	0.005–0.01% (45–90 g./ton)
Bacitracin	4–50 g./ton
Bacitracin methylene disalicylate	4–50 g./ton
Bacitracin zinc	4–50 g./ton
Bambermycins	1–2 g./ton
Chlortetracycline	10–50 g./ton
Erythromycin	4.6–18.5 g./ton
Furazolidone	0.00083–0.0011% (7.5–10 g./ton)
Lincomycin	2–4 g./ton
Oleandomycin	1–2 g./ton of complete poultry ration

Oxytetracycline	5–7.5 g./ton
Penicillin	2.4–50 g./ton
Roxarsone	0.0025–0.005%
Tylosin	4–50 g./ton

Pigmentation	Use Level
Arsanilic acid or Sodium arsanilate	0.005–0.01% (45–90 g./ton)
Roxarsone	0.0025–0.005%

Stomach Appetizer	Use Level
Bacitran	50 g./ton first 4–6 weeks of egg production; 10 g./ton remainder of laying period

SHEEP

Growth Promotion	Use Level
Chlortetracycline	20–50 g./ton
Oxytetracycline	10–20 g./ton

SWINE

Growth Promotion	Use Level
Arsanilic acid or Sodium arsanilate	0.005–0.01% (45–90 g./ton)
Bacitracin	10–50 g./ton
Bacitracin methylene disalicylate	10–50 g. Bacitracin activity/ton
Bacitracin zinc	10–50 g./ton
Bambermycins	2 g./ton
Carbadox	10–25 g./ton (0.0011–0.00275%)
Chlortetracycline	10–50 g./ton
Erythromycin	10–70 g./ton (starter pigs); 10 g./ton (growing and finishing pigs)
Furazolidone	0.0165% (150 g./ton)
Oleandomycin	5–11.25 g./ton of complete feed
Oxytetracycline	25–50 g./ton in swine 10–30 lbs.; 7.5–10 g./ton, 30–200 lbs.
Penicillin	10–50 g./ton
Roxarsone	0.0025–0.0075%

Tylosin	10–20 g./ton (for increased rate of gain and improved feed efficiency, finisher feeds); 20–40 g./ton (for increased rate of weight gain and improved feed efficiency, grower feeds); 20–100 g./ton (for increased rate of gain and improved feed efficiency, starter and prestarter feeds)
Virginiamycin	5–10 g./ton (0.00055–0.0011%) for increased rate of weight gain in swine 120 lbs. to market weight; 10 g./ton (0.0011%) for increase in rate of gain and improvement of feed efficiency from weaning to 120 lbs. (starter and grower feeds only)

Feed Efficiency	Use Level
Arsanilic acid or Sodium arsanilate	0.005–0.01% (45–90 g./ton)
Bacitracin	10–50 g./ton
Bacitracin methylene disalicylate	10–50 g. Bacitracin activity/ton
Bacitracin zinc	10–50 g./ton
Bambermycins	2–4 g./ton
Carbadox	10–25 g./ton (0.0011–0.00275%)
Chlortetracycline	10–50 g./ton
Erythromycin	10–70 g./ton (starter pigs); 10 g./ton (growing/finishing pigs)
Oleandomycin	5–11.25 g./ton of complete feed
Oxytetracycline	25–50 g./ton in swine 10–30 lbs.; 7.5–10 g./ton in swine 30–200 lbs.
Penicillin	10–50 g./ton
Roxarsone	0.0025–0.0075%
Tylosin	10–20 g./ton (for increased rate of gain and improved feed efficiency, finisher feeds); 20–40 g./ton (for increased rate of weight gain and improved feed efficiency, grower feeds); 20–100 g./ton (for increased rate of gain and improved feed efficiency, starter and prestarter feeds)

Virginiamycin	5 g./ton (0.00055%) for improved feed efficiency in swine 120 lbs. to market weight
Litter Production Efficiency	Use Level
Dichlorvos	334–500 g./ton (0.0366–0.0550%) for pregnant swine; mix into a gestation feed to provide 1,000 mg./head/day during last 30 days of gestation

Krider has estimated that each dollar spent for medicated animal feed has yielded five to eight dollars in increased returns to producers.[16]

According to the Animal Health Institute (AHI), the total value of animal-drug sales in the United States during 1980–81 was close to $2 billion. The 1981 figure represents a slight decline from 1980, but when it is compared with 1968 sales, of $411 million, it is evident that animal-drug sales have increased more than fourfold over a fourteen year period.

Antibacterials accounted for a significant share of all animal health-product sales. The 1981 sales of antibacterials as feed additives and pharmaceuticals totaled $438.6 million, or 24 percent of the total health-product market. Almost 59 percent of these sales were as feed additives.[17]

The poultry industry was the prime user of feed additives. Producers purchased a little over 60 percent of the total sales in dollars. Broiler and table-egg producers each accounted for about 45 percent of sales; turkey producers accounted for 10 percent; the remainder of the feed-additive sales were about evenly divided between swine producers and cattle, dairy, and sheep producers.[18]

The AHI reports that farm-animal health-product sales topped $2 billion in 1983, up seven percent from 1982. Feed-additive sales were $1.109 billion—$270.9 million for anti-

TABLE 2

(All figures are in millions of dollars)

Product Class	1981 Preliminary	1980 Revised
Pharmaceuticals*		
Antibacterials	$196.7	$195.9
All others†	398.8	361.3
	$595.5	$557.2
Biologicals**		
Poultry	$ 23.6	$ 26.6
Large animals	78.3	76.0
Small animals	52.3	50.7
	154.2	153.3
Feed Additives		
Antibacterials	$241.9	$241.0
Nutritional	634.0	714.8
All others††	202.2	195.3
	$1,078.1	$1,151.1
TOTAL	$1,827.8	$1,861.7

Source: Animal Health Institute News, 1981 Sales Data.
*Drugs used for disease treatment and control.
†Includes wormers, insecticides, and coccidiostats.
**Vaccines and bacterins.
††Includes hormones and antioxidants.

biotics alone. Total revenues on antibiotics used for therapeutic purposes were $209.7 million and other therapeutic drugs topped $555 million. Sales of biologicals totaled $103.6 million for livestock and poultry use.

The artificial diets we feed to our farm animals consist largely of grains; indeed, the amount of grain currently fed to livestock is approximately ten times the amount consumed by the U.S. population.[19] An estimated 33–50 percent of the world grain supply is fed to livestock, along with 60–70 percent of the high-protein oilseed crop. In the United States,

90–95 percent of the soybean crop, 30 percent of the wheat crop, and 80–90 percent of other cereal grains are fed to livestock.[20]. About 85 percent of all U.S. agricultural land is devoted to the production of animal feed.[21] Grass-feeding of livestock could save 135 million metric tons of grain per year.[22]

The demand for grain (especially grain and soybeans) for meat production has encouraged a trend among farmers to concentrate their production on fewer and fewer types of crops—those with a ready market in animal farming—and to depend on a narrower and narrower range of hybrid, often chemically treated seed. Tilling and crop rotation are minimal or nonexistent on many farms, having been replaced by hazardous herbicide treatment to control weeds and soil erosion. We thus see high-tech farming of cropland as well as animals, which is as ecologically unsafe as the latter is inhumane.

Under the same economic considerations that control the animal farmer—greater productivity in less time and space—the grain farmer too has become heavily dependent on artificial fertilizers and pesticides.

> In 1980, 11.4 million tons of nitrogen, 5.39 million tons of phosphorus and 6.16 million tons of potassium were used to enrich depleted soils—an equivalent of 250 pounds per person, without which, under present cropping practices, yields would drop 50 percent. Resources of these fertilizers are finite: domestic supplies of phosphorus may be exhausted by the year 2000.[23]

The price of fertilizers increased by three hundred percent between 1970 and 1980, and pesticide use is up fifteen hundred percent.

Animal and vegetable byproducts that could be composted and fed back into the land are put into pet foods and farm-animal feeds. For example, poultry manure, rendered remains of animals (such as bone meal), and surplus agricultural produce (such as oranges and pineapples) are given to beef cattle as a cheap source of feed. And increasingly, the valuable manure of farm animals is regarded as a liquid waste to be biodegraded in holding pits—some is actually fed back to the animals—rather than being transformed into nourishment for the fields.

The farming methods we have been looking at are not, as might be suspected, practiced on only a handful of very large, corporate-owned farms; they are pervasive throughout American agriculture. A government study published in 1981 categorizes American farms according to their yearly sales:[24]

TABLE 3

	% of total farms	% of total sales
Rural farm residences *(sales under $5,000)*	44.4	2
Small farms *(sales of $5,000–40,000)*	33.6	16.4
Medium farms *(sales of $40,000–200,000)*	19.6	42.2
Large farms *(sales over $200,000)*	2.4	39.4

Today's 50,000 largest farms account for 30 percent of all farm production. Of the very large farms, only 0.2 percent are truly gargantuan, with sales over $10 million. Nevertheless, 20 percent of the nation's 3 million farms produce 80 percent of our food and fiber. A top 7 percent of the largest farms receive 50 percent of the total cash receipts:[25] The typical commercial farm has sales of approximately $100,000 and assets of $1 million; there are about 600,000 such farms.

In 1983, according to the U.S. Department of Agriculture (USDA), the following numbers of farm animals were raised:

Beef cattle (from 13 major states)	25.7	million
Dairy cows	11.1	million
Lambs	8.2	million
Swine	87.2	million
Turkeys	170.0	million
Broiler chickens	4.18	billion
Laying hens	276.0	million

Concern was expressed a decade or so ago that U.S. agriculture, including food processing and retailing, would soon be controlled by a handful of wealthy corporations. In fact, actual production is still mainly in the hands of farmers, and the retailing and distribution of food, to a much lesser extent, is in the hands of smaller private enterprises. Monopolistic control of the latter may soon be forthcoming, but not of the former. Corporations are learning that it is better to leave farming to farmers, rather than to managers skilled in technology and economics rather than land and livestock. (For their part, farmers are acquiring these managerial skills anyway.) Only an estimated three percent of farms are actually owned by corporations, and many of these are family-owned companies. Corporate control of agriculture is thus effected more by contract farming, and monopoly of off-farm inputs (seeds, machinery, fertilizers, etc.) and off-farm wholesale–retail distribution outlets.

Fifteen companies account for 60 percent of all inputs to farm production.[26] Many seed companies have been acquired by large corporations such as Upjohn, International Telephone and Telegraph, Union Carbide, and Shell/Olin. A monopoly of world seed stock is evolving, along with patent protection. Dekalb, Pioneer, Ciba-Geigy, and Sandoz control some two-thirds of all seed-corn sales. (This trend is causing a reduction in the genetic diversity of plant crops, which reduces adaptability and disease resistance. A similar trend is taking place with various breeds of farm animals.[27]

Diversified, often multinational conglomerates are involved in food production. ITT turns out Hostess Twinkies, Wonder Bread, and Gwaltney's chicken bologna; Greyhound Bus Corporation owns Armour and Company meat packers and Ralston Purina, which markets animal feeds, Chicken-of-the-Sea tuna, and fast food via its Jack-in-the-Box restaurants. Thus, an ever-increasing percentage of American agriculture—from seed to semen to the processing and distribution of farm produce—is controlled by major processors either through contracts with producers or total vertical integration.* The trend is such that the nation's 50 largest food-manufacturing firms accounted for almost 64 percent of the food industries' total assets in 1978, rising from 42 percent in 1963. Of 32,500 food processors in the United States, a mere 100 account for 71 percent of all profits. One company now makes 90 percent of all canned soups, and 4 firms make 91 percent of the country's cereal. Less than 21 percent of all food stores in the United States are supermarkets, yet they accounted for 77 percent of all sales in 1980.[28]

Many corporations now control the major farm inputs, such as seed stock, fertilizer (from coal gas), pesticides (from petroleum), animal feed, veterinary drugs and vaccines, genetic-foundation breeding stock and the new patented hybrid

*Vertical integration refers to the arrangement whereby input suppliers and processors own the farm that uses or supplies their product; thus they control two or more stages of food production. Already, according to the Agribusiness Accountability Project, Washington, D.C., nearly 25 percent of U.S. agriculture is vertically integrated: 94 percent of all vegetable processing, 85 percent of all citrus fruit, 97 percent of all broiler chicken, and 80 percent of all seed-crop production is organized in this way.

strains, as well as building and equipment manufacture.

The rubric "agribusiness" is applied to the wholesale food industry; the petrochemical industry (which manufactures pesticides and fertilizers, and which now monopolizes much of the world's agricultural seed stock); equipment manufacturers; the grain, livestock-feed, and pet-food industries; the pharmaceutical industry; and large corporate farms and contract growers. Agribusiness is an $80-billion-a-year endeavor; pet foods alone account for $4.25 billion.

A multimillion-dollar "hog hotel" in the Parma Township of Jackson County, Michigan, was scheduled for construction in the summer of 1984, despite public protests about odor problems and inhumane conditions for the hogs. The developers, Sand Livestock of Columbus, Nebraska, claim that their pig factory will raise local corn prices, since they will need some 7 million pounds (110,000 bushels) a year to feed the hogs. Three or four local people will be hired to oversee the annual production of nine thousand hogs.

On paper—and to bankers, investors, and economists who can only focus upon immediate, short-term profits, these hog hotels are a boon. And they are a nice tax shelter.

How firm the promise is of increased grain profits for local farmers remains to be seen. As smaller hog farmers in the region are forced out of business by such hog factories that have the economic advantages of size or scale, Sand Livestock will inevitably gain a monopolistic hold. It will be one more step toward corporate integration of hog farming, but with a subtle twist: in the broiler chicken industry, it is the producers who are the corporate peons (see pp. 38–39). But with the new "hog hotel" industry, local producers will be bankrupted, and it is the corn producers who will be contracted to serve the hog factories. And I must emphasize that these "factories" are not in any way like "hotels": they are stressful to the animals, cause unnecessary suffering, increase disease susceptibility, and have been recognized as a significant occupational health hazard to people who work in them.

The only way that large hog hotels like those of Sand Livestock can stay in business is through regional monopoly and a favorable, yet inequitable tax system. This may be the wave of the future for the hog industry, but it should be challenged and opposed, since one of the keys to a sustainable, regenerative agriculture is diversity of farming practices, which the monopolistic trend toward larger and larger hog factories helps eliminate. Aside from the inevitable demise of smaller hog producers when such large, capital-intensive operations come into their area, there are residual problems of local unemployment; ineffective manure (slurry) disposal and utilization; increased risk of epidemic diseases in highly confined, overcrowded animals; and operator health problems (especially respiratory disorders) from working in high-confinement buildings. Land in a wide radius of the "hog palace" will drop in value because of intense and pervasive odor problems.

2

The Economic Dimension

MORE than ever, farmers today are faced with cash-flow problems and extreme variations in net income that add up to a "boom or bust" situation which is becoming more intense each year. Declines in land values and large crop surpluses relative to demand have contributed to even lower returns.

While supermarket food prices have risen by almost forty percent since 1979, farmers' net income has fallen by fifty percent and more. In the second quarter of 1985, the gap between the retail price of beef and what the farmer received grew to a record $1.093 per pound. Between January and June, average retail prices declined three percent, while the prices paid to producers dropped twelve percent. Cattle feeders, who buy the animals and fatten them in feedlot pens, have been hardest hit. And, because of rising transportation and labor costs, there is a long-term tendency toward a widening of the farm-to-retail price spread.[1]

Many farmers are in a financial bind because land prices have fallen. In the 1970s, they were very high and were used to borrow heavily from banks against the assessed value of the property. Some farmers went too deeply into debt; many today have interest rates that make up twenty-five percent of their monthly expenditures. In order to keep up monthly mortgage payments, they must maintain high production and stocking rates. Others expanded their operations only to find that increased production depressed prices on various goods.

Costs of production have increased at a faster rate than profits, so farmers have been forced to cut labor costs, expand their farms, buy more equipment, and go still deeper into debt. This economic treadmill is aggravated by high crop yields and low market prices.

Competition between producers of beef, pork, and poultry, which undercuts their profit margins, forces them to adopt highly intensive husbandry systems that require a considerable initial capital investment. Narrow profit margins necessitate the use of drugs to cut feed costs and to control diseases because of stressfully high stocking densities. Systems of mass production, based on economies of scale, are wrongly justified on economic grounds: the argument runs that if prices were not kept low in the production of beef or pork, consumers would switch to poultry and cheese. One of the consequences of such competitive intensification of livestock production and low price ceilings is that the profit margin of the independent producers is decreasing as feed and other production costs are increasing. This opens the door for corporate monopoly of agriculture, as independent farmers are priced out of business or forced into corporate peonage as contract growers.

A major factor affecting the trend toward factory farming is the present complex federal support system of marketing orders, price supports, farm subsidies, loans, and so forth.

Marketing orders are agreements by the producers of a given commodity that they will ship only that part of their crop that meets a certain size or grade. The market will not, therefore, be flooded with "junk" that depresses the price of the commodity. Marketing orders are enacted by enabling legislation at the state or federal level, and are supported financially by the producers, who assess themselves for the costs of administration, research, and advertising. The net effect of marketing orders often harms the smaller independent producer (who lacks diversified investments and has low profit margins) and to prevent consumers from enjoying low food prices of certain commodities when crop yields are high.

For instance, in 1985 some two billion navel oranges were left to rot in California fields. Significantly, earlier in the same year, after frost damaged the Florida orange harvest, the price of oranges rose some $2.00 above parity because of scarcity. The USDA suspended the "prorate" volume restrictions on the sale of these oranges, and for the first time in decades they were traded in a competitive market. Within a month the price of oranges fell nearly one-fifth and enabled consumers to purchase more oranges at a lower price. This temporary deregulation helped many independent growers make a significant profit. It was opposed by agribusiness: the Sunkist-led cartel lobbied hard to bring back the marketing-order controls.

Tax policies have a profound—and yet, until recently, virtually ignored—adverse impact on American agriculture. Tax laws have been designed that encourage more investments by subsidizing capital investments, and credit policies that encourage expansion. The laws tend to encourage substitution of capital for labor—machinery over people—so that large, mechanized farms get better tax breaks. Smaller operators may enjoy tax breaks but not realize that these same breaks and shelters are helping the "big boys" put them out of business by overproducing and underselling, which they cannot afford to do because their profit margins are too low and they do not have the same diversified investments and capital on hand. And the bigger the farm, the more money it receives from the government under its commodity price-support programs.

The 1978 tax policy made "hog hotels" (confinement buildings; see p. 24–25) eligible for tax credit. The "hotels" and dairy-cow barns are defined as equipment, not buildings. The federal government pays ten percent of building costs and allows depreciation writeoff within five years. This led to a rash of pig factories being constructed with the attendant overproduction and lower prices for the farmer.*

Significantly, some pig farmers are realizing that hog palaces swallow profits. Robert Bryant, a veterinarian and hog producer, presents clear evidence that American hog farming is overcapitalized and that producers put up buildings they can't afford. Then, with high interest rates, they think the solution is in further expansion: more buildings instead of paying off building loans.[2] The tempting tax loophole described above then becomes a constricting noose for the farmer with no off-farm investments and businesses.

Some swine producers now kill their sows after they have had only one litter (while in fact they are more productive with subsequent litters), because the animals can be written off as a capital loss after one year.[3]

A Kentucky hog farmer wrote to this author about the inconsistencies in government policy:

> I went to FmHA [Farmers Home Administration] for an operating loan but was denied because I did not have enough equity. This did not hurt too bad until I learned that FmHA would loan a row crop farmer money to buy seed, fertilizer

*As of May 1984, Sand Livestock had some two hundred hog superfactories in various midwestern states; the entire network is built upon the benefits to private, off-farm investors of tax shelters that the current agricultural tax structure permits.

> and chemicals to plant a crop with the loan secured by the crop he intended to plant, but they will not lend me money to buy feed for my pigs with the loan secured by the pigs I intend to grow.

The Farmers Home Administration, set up to provide emergency loans to farmers, in 1981 refused to give out loans even though they hadn't spent the $600 million that Congress had allocated for the program. At the same time, the Office of Management and Budget decreed that only about $50 million of the $600 million could be used for direct loans to farmers. The rest, OMB ruled, would be used to guarantee loans that the government expected banks to provide to farmers in need.

In 1982, holding to its free-market ideology, the Reagan Administration did not pay farmers to cut production. When this proved a disaster, the Payment-in-Kind (PIK) program* was promulgated, along with another program designed to encourage farmers to store their own excess grain on the farm. This program was intended to reduce federal costs of storing surplus grain bought from the farmers, but instead it enabled farmers to make more money on new crops even when world and domestic market prices were down.

*PIK encouraged farmers to take a certain amount of their land out of production; the crop they would have grown on it would be sent to them by the government out of its surpluses. The intent was both to cut production and to reduce government storage. In actuality, farmers took out their least productive lands. So a farmer might report that he had taken out of production twenty-five percent of his land, say, and receive from the government twenty-five percent of his estimated yield, but in reality that land had furnished only ten percent of his yield. The government thus dramatically overpaid the farmer. The extra produce he had available to sell flooded the market, drastically lowering prices. In addition, many farmers complained that the surplus produce the government sent them was of inferior quality, much of it poisoned by the fumigant EDB.

American food production is boosted for political purposes: food is a political weapon, as two recent secretaries of agriculture have stated, coining the term "agripower." We overproduce, and the excess is traded to enable us to import essential raw materials and less essential consumables, which Americans can buy more easily than natives of countries where food purchases use up most of the income. Vast tonnages of grains and beef are exported to Japan and OPEC countries, for example, primarily for agribusiness profits but also for political, economic, and military purposes. (In this market, the Soviet Union, South America, and Australia are our major competitors.) In essence, our major industry—agriculture—is being used to rescue our failing economy and bolster other industries that have fallen behind foreign competitors.

> Export agriculture has become so vital to the American farmer that the crops from about two of every five acres he harvests now are sold abroad. An astounding 65 percent of the wheat, 55 percent of the soybeans and 35 percent of the coarse grains—corn, sorghum, barley—produced in this country go to buyers overseas.[4]

Exports reached a peak of $43.8 billion in the 1980–81 fiscal year, and are expected to amount to $32 billion in 1984–85. We exported an estimated $4.5 billion worth of livestock and poultry products (including hides) in 1982.

Eighty-five percent of food exports go to developed countries to satisfy their addiction to a Western high-meat diet, while food aid has declined, much going to underdeveloped countries such as Costa Rica, El Salvador, and the Dominican Republic to fatten beef for export back to the United States. We import 100,000 tons of beef annually from Central America. Eight-five percent of all grains exported are not for human consumption, but are used to feed animals, which only the rich in both poor and more affluent nations can afford.

The emphasis of the U.S. government on grain exports, especially with the Soviet Union, has led to confusion and dislocations in American farming and has undoubtedly helped push thousands of farmers toward insolvency. Export markets are uncertain, especially if the Soviets or Western Europe have good wheat harvests.* President Carter's partial embargo of grain sales to the U.S.S.R. in 1980 disrupted the market and resulted in huge domestic grain surpluses.

The federally subsidized land-grant universities and their regional county extension services have been co-opted by agribusiness, through research grants, endowed chairs, student scholarships, et cetera. The agricultural-research budget of the University of California is some $60 million a year, fifteen percent of which comes from private industry and commodity groups. It was discovered in 1982 that the USDA had instigated a formal policy of running loyalty checks on scientific advisors to ensure that their philosophical views were compatible with agribusiness interests,[5] a policy that was immediately terminated by Secretary of Agriculture, John R. Block, once this knowledge became public.

Such federal policies are supported by the agribusiness lobbies. Between 1979 and 1983, dairy organizations contributed $1.8 million to the campaigns of various Congressmen.

The foremost organization that protects the agro-industry is the Council for Agricultural Science and Technology

*European governments have recently given subsidies and price supports to their own farmers, which caused U.S. agribusiness to cry "foul" and threaten to dump wheat surpluses in Egypt and other countries so as to undersell and disrupt the European Economic Community's agricultural markets. Such action harms indigenous farmers by undercutting their own prices.

(CAST), which has a roster of scientific experts who can quickly dispel public concern over any facet of factory farming and agriculture in general. CAST's financial support comes not only from its many member societies, but from agricultural trade associations and drug, fertilizer, and pesticide manufacturers such as Eli Lilly, Mobil Chemical, Monsanto, Dow Chemical, Shell Chemical, Ciba-Geigy, and American Cyanamid. There can be no question of where CAST stands:

> Most recently, CAST's proindustry bias was demonstrated in a dispute with six scientists it had hired to study and report on the use of antibiotics in animal feeds. The scientists' report stated that the practice increases bacterial resistance to antibiotics; CAST reworded their statement in its summary to indicate that the use of antibiotics "might" have that effect. The six scientists quit the study panel in protest. One reportedly believes that the whole purpose of CAST's study project was to produce a document that would be used to counter the Food and Drug Administration's case for a ban on the use of antibiotics in animal feed.[6]

The American Farm Bureau Federation (AFBF), which claims to represent the interests of the family farmer and to be "the Voice of Agriculture," is primarily a front for agribusiness and has an influential voice in many Congressional hearings on agricultural matters. In hearings before the House Agricultural Subcommittee to rewrite the federal Insecticide, Fungicide, and Rodenticide Act, which would benefit the chemical industry, the AFBF favored the chemical industry's redrafting of the act, which would have dropped the farm worker's right to sue in case of pesticide injury and omitted other proposed protections on the field use of dangerous chemicals.[7] Through its network of regional offices, the AFBF exercises a significant influence on local politics and education.

In its 1981 policy booklet, in which it condemns equally homosexuality and the Occupational Safety and Health Act, the Equal Rights Amendment and the Environmental Protection Agency, the AFBF asks that "consideration be given to a risk-benefit ratio" in the establishment of "safe" tolerances for carcinogenic residues in food and that restrictions on banned pesticides be eased under emergency conditions.

> Environmental regulations, whether by air quality standards, water standards, noise standards or visual standards, should recognize the essential nature of efficient utilization of organic matter, pesticides and fertilizers as a basic and natural part of agricultural production.
>
> Normal agricultural practices should be exempt from environmental regulations.[8]

Banking policies represent another factor encouraging the trend toward factory farming. Banks favor intensive systems over more traditional systems or even renovation of less "modern" high-tech automated systems, and this is where they put their loans. Farmers are co-opted into adopting such systems or going out of business.

They are further encouraged to adopt such systems by agribusiness magazines, trade-association promotions and salesmen, and by university and agribusiness-allied animal-science technologists. The many farm-support industries are wholly dependent upon the farmer for their existence. While some farmers, such as broiler producers, are under corporate peonage as contract growers,* others have been effectively converted to the belief that they cannot farm or make any profit if they do not build highly automated confinement buildings. And, committed to a virtually inflexible system, they go heavily into debt in the process of setting up and

*Contract growers usually have all aspects of the animals' rearing, including details of feed and drugs, dictated to them by the corporation with whom they contract.

maintaining it. Nor can they raise stock, the drug companies tell them, without using antibiotics and other drugs in the feed.

Factory farming entails considerable financial cost to both the farmer who practices it and the more traditional farmer who must compete with it.

Economies of scale mean that the farmer measures his profits *per acre* or *per herd*. A 1981 CAST report on scientific aspects of the welfare of food animals guardedly admits:

> One must recognize the existence of some degree of conflict or trade-off between animal welfare and human welfare because the combination of conditions that leads to the maximum profitability of an animal-production operation involving many animals is not necessarily the same as the combination of conditions that leads to the maximum welfare of the animals individually.[9]

In 1980, a midwestern farmer had several dozen pigs stolen. He didn't notice when they were stolen, because he only went into the "factory" about once a week. The building is so automated that he only needed to check his pigs every few days. But automated farming won't tell him when a pig is sick. Another pig farmer boasted in *Hog Farm Management Magazine* that he has such an efficient system that his pregnant sows don't need to be fed for ninety days. He allows them to find what they can in the manure waste pits under the slatted-floor pens where young pigs are being fattened for slaughter.

The economic structure of industrialized animal farming is such that there is no financial incentive to maximize individual welfare and productivity: both are sacrificed to increase efficiency and overall profits. Scientists have published research findings on optimal stocking densities which show that most commercial stocking rates clearly reduce individual production and overall health and welfare. However, such findings are not applicable to industrialized animal farming, where maximal profits depend less upon optimal individual care, welfare, and productivity and far more upon increased efficiencies of scale.

As new industrialized systems become established, new norms are set in terms of acceptable standards of animal care, stocking densities, feed-conversion ratios, disease incidence, and kinds of disease. A high incidence of lameness from a variety of causes and pneumonia are the accepted norm in swine-confinement operations, for example, as is dependence upon drugs, vaccines, and technology (ventilation and manure-disposal systems, etc.) to control those diseases that animals become more susceptible to when raised intensively.

Veterinary care shifts from a focus on individual treatment to the concept of herd-health maintenance. Animals in need of individual care are culled if they do not recover spontaneously or respond to minimal (and often improper and inadequate) treatment by nonveterinarians, even though, as studies have shown,[10] this practice can jeopardize overall herd health. The turnover rate of stock increases, since the longevity and welfare of the individual have a lowly niche in the hierarchy of economic concern.

Nevertheless, when drugs and veterinary expenses are totaled, the farmer's outlays have not diminished:

> Total farm production cost increased 53.01% from 1972 to 1976. At the same time, however, total farm animal health cost increased 84.52%. Regarding costs per farm, total production cost rose 71.51% whereas total animal health cost increased as much as 91.07%.
>
> In conclusion, it is clear that while farmers' total production costs have increased considerably in recent years, farmers' expenditures for animal health care have increased at an even faster pace.[11]

Estimated veterinary-care expenditures for the major classes of farm animals in 1981 were:[12]

Beef cattle	$386.0 million
Dairy cows	$306.0 million
Swine	$68.0 million
Poultry	$38.0 million
Sheep	$12.7 million
TOTAL	$810.7 million

Livestock health costs to farmers increase disproportionately with increasing scale of production.[13] We have reached the point where some farmers cannot even afford to call in a veterinarian to treat their stock.[14] Consequently, the veterinary profession is beginning to hurt financially. Yet some veterinary associations mistakenly perceive the animal-welfare movement as an economic threat.

Heavy losses of animals and vegetables are accepted, so long as the overall operation shows a profit. An estimated twenty to thirty percent of all poultry and livestock die before they reach mature slaughter age, at a loss to consumers of $4.6 billion per year (a figure reported in the *Federal Register* in 1981 relative to a proposal to use even greater amounts of drugs in animal production and to reduce regulatory restrictions and public opposition to drug-dependent farming). A further estimated $1 billion is lost annually from disease, stress, shrinkage, bruising, and death associated with livestock transportation (primarily by road, farm animals have no protection under the existing livestock-transportation law, which makes provisions only for proper care and handling by rail car).

Factory farming is, as has been noted, capital-intensive farming. Economies of scale imply that farmers must expand to create that scale; thus we have the huge loans they have had to take on to purchase more land and more and fancier equipment.

The extensive purchase of seed stock from outside sources—usually large corporations—is another considerable expense for the grain farmer, as is feed for the animal farmer who no longer grows his own. Other large expenses include animal drugs, fertilizers, and pesticides. Professor David Pimental estimates that about $2.5 billion of pesticides is applied to cropland (much of the rest going to forestry and aquatic-resource management and in spraying trees and roadsides in urban and suburban areas).[15] The use of herbicides alone has increased from an estimated 207 million pounds in 1971 to an astronomical 451 million pounds in 1982.

Large-scale, intensive operations have evolved wherein there is reduced human contact with animals and thus a reduction in labor costs. Equipment manufacturers and some animal scientists favor such capital-intensive systems in spite of evidence that some of these systems actually require *more* labor (and more than twice as much veterinary care and drugs) for each unit of food produced than smaller and less intensive operations.[16] Dr. Robert Bryant dispels the myth that high-capital, low-labor facilities are more profitable than labor-intensive units: "low-labor" confinement buildings require a lot of work in repairs and maintenance.[17]

Farmers caught in the bind are sometimes forced to become contract growers, raising stock or crops for a large corporation, and so are reduced to peonage on their own land. The difficulties contract poultry growers face have been well documented by Hope Shand. In a review of the broiler-chicken industry in the South, she writes that

> The contract system allows agribusiness to treat farmers as employees—without any obligation to pay them as employees or provide them with job security. As long as the farmers are under contract they are considered independent operators. The company thus has no obligation to provide health insur-

ance, workers' compensation, paid vacation and so on. And the industry wants to keep it that way.[18]

Many independent smaller operators are forced to rely on off-farm jobs to supplement their incomes (they can also subsidize their farms by using the same tax code that benefits larger conglomerates and which *in toto* is estimated to cost the U.S. Treasury $2.6 billion between 1985 and 1987).[19] Those who lose their farms move to the cities and are added to the lists of the unemployed.

Confinement buildings require large amounts of electricity, so fuel bills are an important part of the animal farmer's overhead, as they are of the farmer who artificially irrigates his fields.

In light of all these costs and uncertain markets, many farmers have not been able to keep up with their loan payments, and farm bankruptcies have escalated. A 1979 study reported that nearly two thousand farms (and three hundred rural businesses) go out of business each week.[20] During the 1970s, fifty-seven percent of all black farmers went out of business, reflecting a major loss of black-owned land, according to a 1982 U.S. Civil Rights Commission report.

Dr. Barry Commoner, in *The Poverty of Power,* acknowledges that between 1950 and 1970, American agriculture impressively increased its productivity: corn production per acre tripled; broiler chickens gained nearly 50 percent more weight from their feed; egg production increased by 25 percent; overall farm production increased by 40 percent. However, during this same period, real farm income decreased from about $18 billion in 1950 to $13 billion in 1971. The number of farms decreased by 50 percent, while income per farm rose by 46 percent (the average increase in the family income of all U.S. families in that period was 76 percent.) Meanwhile, the total mortgage debt of American farms rose from about $8 billion in 1950 to $25 billion in 1971. Net farm income declined during this twenty-year period because farms became larger, more specialized, and increasingly dependent upon fossil fuels and the new agribusiness technology of chemicals and machinery.

Commoner observes that

> One can almost admire the enterprise and clever salesmanship of the petrochemical industry. Somehow it has managed to convince the farmer that he should give up the free solar energy that drives the natural cycles and, instead, buy the needed energy—in the form of fertilizer and fuel—from the petrochemical industry. Not content with that commercial coup, these industrial giants have completed their conquest of the farmer by going into competition with what the farm produces. They have introduced into the market a series of competing synthetics: synthetic fiber, which competes with cotton and wool; detergents, which compete with soap made of natural oils and fat; plastics, which compete with wood; and pesticides that compete with birds and ladybugs, which used to be free.
>
> The giant corporations have made a colony out of rural America.[21]

The kind of farming we have been practicing also elicits considerable indirect costs from the farmer and farm worker.

There is evidence that certain forms of cancer and other related health problems are more prevalent in farmers than in the rest of the rural community. This is because it is the farmers who handle pesticides and other chemicals that have been shown, in laboratory animals, to cause cancer, sterility, birth defects, and a host of other disorders, including reduced resistance to disease. (Similar disorders have been found in people who work in the factories that manufacture these chemicals.)* The serious occupational health hazards

*The Velsicol Chemical Company of Houston, Texas, was charged by

to people working in total-confinement swine operations is now an additional recognized problem, where dust and toxic gases cause acute and chronic respiratory difficulties.[22]*

Until recently, capital-intensive farming that favors an increasingly select few has been further weighted by the inflated paper value of agricultural land. During such periods, farmers can hedge inflation and still stay in business by securing loans that use the value of their land as collateral. But this practice makes it virtually impossible for a young farmer to start his own business: he simply can't afford to buy the land. On the other hand, when inflation decreases and land prices plummet, the high interest rates on loans lead to the demise of thousands of independent farmers. Some thirty

---

the federal government (the National Institute for Occupational Safety and Health) with the withholding from federal inspectors of vital information about health problems among their workers in their Leptophos plant. The company has also been indicted by a federal grand jury for concealing laboratory test results from the Environmental Protection Agency which indicated that Leptophos may be carcinogenic. Another pesticide, DBCP, which is now banned in the United States, has made some workers at the Dow Chemical and Occidental Petroleum companies plant sterile.

*Workers in slaughter plants suffer the pressures of extremely tedious work (such as cutting out one part of each animal that passes by on a conveyor). This work has to be done at high speed—inhumanly high speed since, according to the U.S. Department of Labor, almost one fourth of the workforce in the poultry industry suffers each year from industrial injuries. Speed means more profits for slaughterhouses, and, bowing to the meat packing industry's demands, the rate of birds inspected per minute by the USDA was raised from 45 to 70 under the Carter Administration, and then to 85 under Reagan; a speedup of cattle- and pig-carcass inspections has also been proposed.

Critics contend that it is not humanly possible for federal meat inspectors, who complain of "line hypnosis," to do an adequate job under these new procedures, and that this will mean that more animals and animal parts unfit for human consumption will be on our tables. For details, see Kathleen Hughes, *Return to the Jungle: How the Reagan Administration Is Imperiling the Nation's Meat and Poultry Inspection Program* (Washington, D.C.: Center for the Study of Responsive Law, 1983); and Kathleen Hughes, *New York Times*, March 2, 1983.

percent of all pork producers went out of business between 1980 and 1982.

So a system of land-leasing has evolved, wherein larger, wealthier farmers, absentee landowners, and corporations lease out some of their acreage to poorer and younger farmers.* This system has been likened to the tenant farming and sharecropping exploitation of southern blacks that was widespread a few decades ago. Alternately, the young farmer may become a contract manager for a corporate farm or absentee landlord.

USDA projections indicate that, by the turn of the century, today's 50,000 largest farms, which they state now account for thirty percent of all farm production, will produce two-thirds of the total. The process of farm enlargement and intensification is an economic-technological-chemical treadmill which sacrifices rural communities and traditional farming knowledge attuned to the local nuances of climate, soils, and seasons (since few farmers' offspring can afford entry into farming) for an agriculture that displaces humanistic, cultural, communal, esthetic, and spiritual values: the agrarian ethos is industrialized. The larger the surrounding factory farms, the lower is the quality of life in the rural community.[24]

Wendell Berry observes:

> The concentration of farmland into larger and larger holdings and fewer and fewer hands—with the consequent increase of overhead, debt, and dependence on machines—is a matter of complex significance, and its agricultural significance cannot be disentangled from its cultural significance.
>
> It forces a profound revolution in the farmer's mind: once his investment in land and machines is large enough, he must forsake the values of husbandry and assume those of finance

*A 1974 survey showed about 1.5 million nonoperator landowners, owning 330.4 million acres of land which is rented to farm operators.[23]

> and technology. Thenceforth his thinking is not determined by agricultural responsibility, but by financial accountability and the capacities of his machines. Where his money comes from becomes less important to him than where it is going. He is caught up in the drift of energy and interest away from the land. Production begins to override maintenance. The economy of money has infiltrated and subverted the economies of nature, energy, and the human spirit. The man himself has become a consumptive machine. . . .
>
> The mind of a good farmer is inseparable from his farm, or, to state it the opposite way: A farm, as a human artifact, is inseparable from the mind that makes and uses it. The two are one. To damage this union—as industrial agriculture now threatens to do irreparably—is to damage human culture at its root.[25]

Americans, we are told, have the lowest food prices in the world. Of the money we do pay for food, only a small percentage goes to the farmer.

Today, the main profits in the livestock industry are reaped by middlemen, processors, and retailers of animal produce. According to one study, supermarket monopolies inflate prices $1 billion annually. Consolidation in the food industry reduces competition and, in 1980, led consumers to pay an estimated extra $18 billion. The food industry spends $7–9 billion on advertising, which costs consumers 3–4 cents for every dollar of food purchased. Nine percent of the consumer's total food expenditures—some $34 billion per year—pays for food and beverage packaging.[26]

The "cheap" food we buy must be paid for twice: once at the market and again when we pay taxes. Via the government, the public pays out billions of dollars in price supports and subsidies, low-interest loans, and crop insurance, as well as for the purchase and storage of surplus produce (especially grain and dairy products).

We also pay billions for the U.S. Department of Agriculture administration, as well as for the research and training done at land-grant colleges, which rarely benefit the family farmer. Various public-relations firms and livestock-industry lobbyists work collectively to secure tax moneys to fund research into livestock production and to promote the consumption of meat, eggs, and dairy products. Additional public expenditures go to the Department of Health, Education, and Welfare and the Environmental Protection Agency (EPA) to monitor pesticide and other chemical residues in our meat, milk, fruit, cereals, vegetables, drinking water, and air, and to research cures for sicknesses related to our diet and polluted environment.

In his weekly radio address of August 17, 1985, President Reagan noted that the federal government has spent nearly $59 billion since 1981 to support the price of farm products, more than three times the amount spent from 1976 to 1980.[27]

The tobacco industry alone is subsidized to the tune of $50 million each year. Six billion dollars was paid out to the dairy industry between 1982 and 1985 for products that cost more billions to keep in storage. Surpluses that are bought and stored by the government may then be sold back to producers (for instance, milk surplus is purchased by veal producers) for a nominal sum.

In 1984, some of the largest western wheat and cotton farms, which also benefit from public-funded irrigation programs (see p. 48), received PIK payments worth over $1 million each.

The Reagan Administration's attempt to reduce overproduction in the dairy industry in the spring of 1984 resulted in some dairy superfarms receiving more than $1 million each for not producing milk. One California dairy received approximately $3 million, while three of the highest subsidy recipients were in Florida, which has no milk surplus and is not part of the cost problem that afflicts the dairy-support program. (Ironically, many Florida dairy farmers were planning to cut back on production because of increased feed-grain prices following the drought of 1983.)

The budget of the United States government for the fiscal year of 1982 includes regulatory controls for meat and poultry inspection, a commodity-purchase service, egg-products inspection, and a voluntary commodity inspection and grading program. The budget outlines the following federally funded inspections:

OBJECT CLASSIFICATION
(in thousands of dollars)

Federally Inspected Establishments	1980 actual	1981 estimate	1982 estimate
Meat and/or Poultry Slaughter Plants	504	513	520
Meat and/or Poultry Processing Plants	5,014	5,035	5,045
Meat and/or Poultry Slaughter & Processing Plants	1,543	1,575	1,610
Egg-products Plants	119	120	120

The 1982 budget includes a laboratory-service section. But note that the object classification is calculated according to "samples analyzed" not "thousands of dollars."

Laboratory Service	1980 actual	1981 estimate	1982 estimate
Food Chemistry	100,000	130,000	151,000
Chemical Residues	34,000	35,000	46,000
Antibiotic Residues	26,000	55,000	69,000

According to the agricultural, rural development, and related agencies' appropriations for 1982, hearings before a subcommittee of the Committee on Appropriations House of

Representatives, Part 4B, the budget on page 47 for the residue program was determined.

In the spring of 1984, millions of chickens had to be destroyed in Pennsylvania and adjacent states (over 11.5 million in Pennsylvania alone), costing taxpayers around $35 million in compensation to farmers. The birds were killed to prevent an outbreak of avian influenza from decimating the American and Canadian poultry industries. The epidemic was principally created by factory-farming practices. The birds were made vulnerable by overcrowding and the stresses of accelerated growth or egg production.

FmHA makes low-interest loans to farmers; the difference between the interest rates charged and prevailing bank rates is, of course, paid by the taxpayer. The taxpayer also pays when the borrower of government-guaranteed loans defaults.

According to the U.S. Treasury, farming schemes amount to only three percent of the total revenues lost to the government via tax shelters such as those already described (p. 29 above). Nevertheless, the tax code will cost $2.6 billion between 1985 and 1987.[28]

Rail and truck transportation amounts to 5.3 percent of the retail cost of food. (Consumers paid almost $16 billion in 1980 for transportation; an average food item travels an estimated 1,300 miles before being eaten.) Taxpayers subsidized the construction of the national Interstate Highway System, which led to the demise of the more efficient rail-transport system. Now 99 percent of all livestock, 88 percent of fresh fruit and vegetables, and 80 percent of dairy and bakery products are carried by road. One fully loaded forty-ton truck does as much damage to the roads as 9,600 automobiles, and there are 4 million trucks in use.[29]

Public funds given to land-grant universities under the Hatch Act amounted to $152.3 million in 1984. A court order

TABLE 4

RESIDUE PROGRAM
*(thousands of dollars per slaughter year)**

	1981 Base	Educational & Cooperative Program	Contamination Response System	Detection	1982 Increases	1982 Budget
Slaughter	7,195/332	1,220/0	150/4	265/6	1,635/6	8,830/338
Import-export	755/35			112/1	112/1	867/36
Laboratory Services	6,149/196	153/4	1,058/3	4,742/27	5,953/34	12,102/230
Compliance		9/0	254/8		263/8	263/8
Eggs-products Inspection	———	———	44/1	151/2	195/3	145/3
TOTAL	14,099/563	1,382/4	1,506/16	5,270/36	8,158/52	22,207/615

*A slaughter year refers to the number of staff years of individual workers. Thus 7,195/332 means $7,195/332 staff years.

was sought in 1984 by the federally financed California Rural Legal Assistance Foundation (CRLA) to require the University of California to assess research proposals for potential adverse social consequences and to undertake only those projects that will benefit the intended beneficiaries—mainly small farmers—of the Hatch Act. CRLA's contention was that the university is violating state and federal law by using public funds for mechanization research (such as a tomato harvester) that benefits agribusiness at the expense of small farmers, farm laborers, and rural communities. They lost their case.

Superfarms in California's Imperial Valley receive tax-subsidized irrigation water (administered by the Interior Department) to boost production as other farms are paid by the Agriculture Department to reduce production. According to the CRLA and the National Resources Defense Council (NRDC), the federal government has paid $1.5 billion in illegal subsidies to large farming interests to keep water prices artificially low in California's Central Valley over the last forty years. A year-long study[30] showed that farmers in the Westlands Water District alone receive an average water subsidy of $500,000 a year from the U.S. Bureau of Reclamation. In Westlands, growers pay an average of $9.45 per acre-foot for water, which the report maintains would cost $97 without a subsidy (Interior Secretary Donald Hodel contests these specific figures). Only $50 million of the $931 million spent on irrigation facilities in the Central Valley has been repaid because of extensions of repayment periods; misuse of "ability to pay" provisions that reduce the growers' obligation to repay their full share of costs; and long-term, fixed-rate water contracts at rates far below cost. These subsidies are scheduled to continue until the year 2030.[31]

An incalculable but not inconsiderable cost that must be

added to the total food bill is what we pay in welfare benefits as small farmers lose their farms and farm workers are replaced by technology: both these groups migrate to the cities and are added to the ranks of the urban poor. The farm labor force has been almost totally liquidated over the past four decades; the company planning the "hog hotel" described on p. 25 above announced that it planned to hire three or four people from the community to help oversee the yearly production of nine thousand hogs.

Residents of states that rely on large migrant forces for harvesting are familiar with the annual news reports that welfare agencies have geared up to assist the families of workers in such areas as food stamps and day care. Thus we are informed that the salaries paid to the harvesters and passed along in our food bills are not the whole story of what we pay.

3

The Health of the Earth and Its People

THE RATE of entropy (defined as the depletion of fossil fuels and other resources) is accelerating. This reduces the number of our future options. The wholesale industrialized exploitation of nature for short-term gain cannot be sustained; it is a major component of what we have termed agricide. Agricide includes escalating soil erosion; the impoverishment and poisoning of the soil; the pollution of lakes, rivers, and oceans from the runoff of topsoil contaminated with chemicals; the destruction of groundwater sources from pollution and overuse; the use of nitrogenous fertilizers that are also implicated in the destruction of the earth's ozone layer; and the development of fast-growing, high-yield hybrid strains of crops and animals that are more susceptible to disease. An estimated eighty-five percent of all U.S. agricultural land is used in the production of animal foods, which in turn is linked with deforestation, destruction of wildlife habitats, extinction of species, loss of soil productivity through mineral depletion and erosion, water pollution and depletion, overgrazing, and desertification.

The USDA estimates that cropland erosion is occurring in the United States at a rate of two billion tons of soil a year.[1] (Other analysts contend that the rate is four to six billion tons.) In 1970, the National Academy of Sciences reported that the nation had lost one-third of its topsoil. Soil erosion and land development together account for an estimated loss of some thirty-four square miles of U.S. agricultural productivity every day.[2]

The same problem is taking place throughout the world. At the present rate of land degradation, one-third of the world's arable land will be desert by the turn of the century.[3] Congress has taken action to help prevent one major source of soil erosion—that by "sodbusters." The sodbuster phenomenon in western states has entailed the plowing and seeding of hundreds of thousands of highly erodible rangeland, motivated by federal farm supports and the increased resale value of such land to investors seeking tax shelters.

It is quite likely that the Department of Agriculture has come up with a solution to soil erosion that is actually worse than the problem they are endeavoring to correct: more farmers are now practicing what is called conservation tillage, which entails less frequent tilling of the land; instead, repeated spraying with herbicides is done to control weeds. One of the main problems with conservation tillage is the further poisoning of groundwater and contamination of crops grown on poisoned land.

T. C. Byerly estimates that, of the 33.5 billion acres of land surface on earth, there are some 3.5 billion acres of arable land, 7.75 billion of pasture and rangeland, 10 billion of forest/woodland, and 12 billion of other forms (desert, mountain range, swamp, etc.). In the United States, Byerly calculates that 70 million acres of land are used to produce agricultural produce for human consumption (including exports of food), while 187.5 million acres are used for producing animal feeds for domestic and foreign use. (Some 2.5 million acres are required for producing beer, wine, and distilled liquors.) He cites a study by the United Nations Food and Agriculture Organization that finds there is at least as much uncultivated land available in the world that is physically suitable for crop production as the total amount that is presently used.[4]

While development reduces the amount of land in farms, other lands are cropped or grazed to take its place. Rich hardwood forests of the lower Mississippi Valley are being systematically cleared for conversion into cropland. Each year we import 100,000 tons of beef from Central America, where tropical rain forests are being destroyed at an accelerating rate to open up grazing land for cattle production. Globally, deforestation is occurring at a rate of 150 million acres a year.[5]

Botswana, the last wilderness region of Africa, like much of the Amazon jungle, is being cleared and turned into a vast cattle ranch. This export trade has been supported by the European Economic Community to provide nineteen thousand tons of beef annually for the convenience and fast-food industries. Thanks to such economic aid, Botswana ranchers have erected eight hundred miles of fencing to keep wild animals from grazing and infecting their cattle with disease. But these fences have disrupted the migration routes of wildebeest and other herbivores in search of water and fresh grazing. In recent dry seasons, tens of thousands of animals have died.[6]

An international boycott of Burger King and other fast-food chains was launched in the spring of 1984 by John Seed, of the Australian Rainforest Information Center, and the radical "deep ecology" Earth First! network. Seed points out that forty percent of the tropical rain forests of Central America have been destroyed, primarily to raise cattle for the cheap-hamburger market in the United States.

With the best intentions at heart, animal scientists, veterinarians, and government agencies have been doing their utmost to help the countries of the Third World develop more productive cattle- and sheep-rearing systems. Yet in many parts of the world, such activities are ecocidal: the wrong species are being put in the wrong places. Cattle in particular are not adapted to semiarid or seasonably arid climates, such as some regions of the United States and the Asian and African continents. They require water, unlike indigenous species such as gazelle and antelope. Cattle are also more destructive of natural vegetation in their grazing patterns, which leads to soil erosion and drought.

The inappropriate application of the technology of livestock production can have calamitous, unforeseen sociopolitical and environmental consequences, as happened in the tragedy of the African Sahel (where Western intervention, aimed at increasing livestock production, resulted in war, famine, and desertification).

Dr. David Hopcraft, a wildlife ecologist in Kenya, dramatically demonstrates what happens when nonindigenous and inappropriate species are introduced into a low-rainfall grassland plains ecosystem:

> Tracking and increasing devastation around water holes reduces not only the amount of grazing available, but also the amount of moisture which is normally passed from the soil to the atmosphere by the transpiration of plants. The air thus becomes drier. The bare earth is now exposed to the sun, thus heating up and killing off the micro-organisms. The dry and heated air rises, reducing further the probability of rainfall.
>
> If it does rain it now does more harm than good, for the water runs, and erodes the lifeless soil. The end result of this process is Desert and Death.[7]

Hopcraft's 20,000-acre wildlife ranch near Nairobi, where he raises wildebeest, gazelle, and other indigenous herbivores for human consumption, is ten times more productive and profitable per acre than neighboring cattle-ranching operations.

Millions of acres of land are being turned into desert each year, over half a million in the Sudan alone. The major cause is overgrazing by cattle. In many African countries, as many as ninety-nine percent of the indigenous wild animals have disappeared forever.

Dr. I. Garth Youngberg, Executive Director of the Institute for Alternative Agriculture, and Agricultural Research Service (USDA) scientists J. F. Parr and R. I. Papendick observe:

> Up until four decades ago, conventional agriculture in the U.S. was, for the most part, beneficial to the support and proliferation of many wildlife species. [However, the emergence of] monoculture grain production along with intensive row cropping, clean tillage cultivation, larger machinery, and heavy applications of chemical fertilizers and pesticides [resulted in] a concomitant decline in the (wildlife) food base, habitat areas, and in turn, the numbers and species of wildlife. . . . The need and desire to preserve and enhance fish and wildlife resources is one of the most important factors accounting for the increased interest in alternative agriculture.[8]

The American Farm Bureau Federation, on the other hand, proposes:

> The National Endangered Species Act of 1973 should be amended to provide that:
>
> Listing a species as endangered shall be upon that basis alone and not on the basis of "rarity";
>
> The law shall not encroach upon economic agricultural or silvicultural practices; . . .
>
> Where there is a conflict, human need for food and energy

> should have priority over the protection of endangered species. . . .
>
> We support the establishment of statewide or interstate contracts designed to administer a predator bounty system. We support continuation of all established predator-control practices, including traps and chemical toxicants, specifically 1080, under federal or state supervision. We support aerial hunting to help control predator numbers.
>
> We urge that the timber wolf be declassified [as an endangered species].[9]

The U.S. cattle industry is beginning to turn back to raising cattle on range and forage because fattening and finishing cattle primarily on grains is becoming too costly. Thus, to cut production costs, this ecologically unsound industry is going to intensify its impact on the natural environment. *Beef* magazine states:

> Grazeable land is America's largest natural resource. One-half of the land area in the United States is covered with forage. That's more than one billion acres, enough to give every cow 200 acres to graze all by herself!
>
> This grazing land is made up of rangeland, natural and improved pastures and forests. From a production standpoint, grazing land accounts for more than half of the meat, a third of the milk and all of the wool produced in this country. In terms of dollars, grazing land generates more income than General Motors, Ford and Chrysler combined. Surprisingly very few livestock producers consider grazing land their major resource.[10]

The article goes on to discuss the efficiencies of controlled grazing—dividing the land into smaller enclosures that are intensively grazed and then rested. But, as we know, fences block the movement of wildlife, denying them access to seasonal grazing areas, water, etc. (This is often deliberate, along with their extermination, in many parts of the world to protect livestock from diseases wild herbivores may transmit.)

While we may find it esthetically pleasing to see green rolling hills covered in sheep, and cattle roaming on the wide open range, we should not be tricked into thinking that these pastoral scenes are natural. They are industrialized landscapes that should be returned to nature.

Cattle- and sheep-ranching industries are subsidized by the public, and private landowners enjoy tax advantages for clearing tropical rain forests and other natural ecosystems for agricultural development as has been done in Hawaii.[11]

This is not to say that natural grazing is undesirable. Professor Calvin Schwabe observes that, of the 37.5 billion acres of the earth's land not covered by ice, only 3.75 billion acres could be cultivated, and most of this is already under plow. Some one percent of the sun's energy reaching the earth is stored in plants. The utilization of this energy for feeding man is possible *only* through the activities of grazing and browsing animals that harvest under their own power and convert—into the highest-quality human food—highly scattered or otherwise inaccessible plant life. Some 7.5 billion acres are usable for this purpose, at a reasonable level of production, worldwide; this is about twice the amount of land that is available for cultivation. Schwabe estimates that sixty percent of the world's animal protein production, about thirty million metric tons, now comes from these nonarable lands. Herbivorous animals, and particularly ruminants such as cattle, "provide the *sole* vehicle by which much of the earth's surface can be exploited for food production."[12] But, as Dr. Hopcraft has shown; it may be more prudent to replace cattle, sheep, and goats with indigenous wild ruminants.

In some areas, the land is saturated with synthetic fertilizers; rivers become polluted with the water runoff from these areas.

> The people of the United States . . . are pouring into the sea, lakes, or rivers and into the underground waters 6 to 12 million pounds of nitrogen, 2 to 4 million pounds of potassium and 75,000 to 3 million pounds of phosphorus per million of adult population annually and this waste we esteem one of the great achievements of our civilization.[13]

Copper now poisons Kenya's soils, after being used for years as a weed killer on coffee plantations.

The widespread use of insecticides results in the non-selective poisoning of both "good" and "bad" insects. Worse, this practice can lead to the development of resistant strains and to the unchecked multiplication of currently harmless but naturally resistant insects, which then become pests because the other insect species and predatory birds that normally keep them in balance have been killed off.

According to Dr. H. M. Caine of the University of California, Santa Cruz, one-third of the American diet depends directly or indirectly on crops pollinated by honey bees and six percent of farm production ($3.5 billion) is at least indirectly dependent on such pollination. But pesticides are now at work reducing the bee population, by an estimated two percent per year. This loss could have serious agricultural consequences. Nevertheless, the AFBF supports the "modification of existing regulations to more easily permit restricted use of previously cancelled pesticides under emergency conditions."[14]

The problems of dealing with the 1.24 billion tons of solid and liquid waste from animals are astronomical. This byproduct is not a natural manure that can be easily recycled onto the fields. Rather, it has high concentrations of drugs, arsenic, and copper, and abnormally high amounts of undigested protein; it is therefore a serious freshwater pollutant. Disposing of manure in oxidation lagoons deprives the fields of nitrogenous fertilizer.

There is also the problem of topsoils that show declining levels of essential nutrients (nitrogen, phosphorus, potassium, etc.), which, in turn, decreases crop production and resistance to disease, impairs the health and productivity of farm animals, and lowers the nutritive value of the crops to consumers.

On the other hand, soil erosion and irrigation can lead to the accumulation of toxic levels of trace elements in lakes located in drainage areas distant from agricultural activities. And in some areas there is natural excess, as with selenium in Southern California, and with aluminum over much of the world. This is being leached from the soil by acid-rain pollution* and now contaminates lakes, rivers, and drinking water. (It has been implicated in Alzheimer's disease, or premature senility.)

Air pollution is a serious threat to agricultural productivity. Sulfur dioxide and nitrous oxide emissions from electrical utilities and automobiles cause harmful acid rain. Nitrous oxide also breaks down into photochemical oxidants, which are toxic to plants and impair photosynthesis. It has been estimated that a reduction in ambient ozone levels of 25 percent would produce nearly $2 billion in benefits, while a 25 percent increase would lead to an additional $2.3 billion in crop losses.†

Waterfowl are now suffering the consequences of im-

*According to Professor Bernhard Ulrich of the University of Göttingen, acid rain leaches essential nutrients such as potassium and calcium from the soil. It also deposits toxic metals such as lead and aluminum and thus prevents soil microorganisms from converting organic debris into fertilizer.

†R. M. Adams, S. A. Hamilton and B. A. McCarl, *The Economic Effects of Ozone on Agriculture* (Corvallis, Oregon: 1984); Environmental Research Laboratory, Environmental Protection Agency.

proper land use in the San Joaquin Kesterson reservoir in California. Selenium (from distant, selenium-rich fields) has built up in plants and fish the birds eat. Forty percent of waterfowl eggs contained dead embryos in 1983, and twenty percent of hatched chicks had deformities: swollen heads, no eyes, legs, wings, etc.[15]

According to Larry Ephron, an agricultural historian and analyst, the German chemist Justus von Liebig, over a century ago, analyzed the ash residue of burned plants and found that the primary components were potassium, nitrogen, and phosphorus. His findings became the basis of using these artificial chemicals as fertilizer. Had he available then modern analytical equipment, he would have been able to identify more than ninety different essential elements in plants that should be returned to the soil. Apparently realizing the hazards of using such a limited range of artificial chemical fertilizers which he had helped stimulate, he wrote that he had "sinned against the wisdom of the Creator."[16] Applying potassium, nitrogen and phosphate certainly boosted crop yields, but this caused serious imbalances in the soil and other essential nutrients were not put back.

The deficiencies in our agricultural soil and the crops we and farm animals eat are so marked that simply putting crushed gravel screenings onto the soil and kiln dust into animal feed will dramatically increase the farmer's yields of grain and meat alike.

Many of the diseases that afflict us, our domestic animals, and our crops and forests are related in part to deficiencies in certain essential trace minerals. Immunosuppression and increased susceptibility to stress, pathogenic organisms (especially viruses), and chemical (natural and industrial) poisons, carcinogens, mutagens and teratogens, and possibly emotional instability* have been linked with trace-mineral deficiencies and imbalances.

*A trace element called lithium (first used in the town of Lithia in

The CBS Evening News on May 21, 1984, showed a clip of Wyoming residents gathering the antlers that had been shed by elk in their government winter reserve. The news item was presented as a piece of humor about rural—indeed wild, pioneer—America and showed an antler auction and a Korean buyer writing out a check for $30,000: ground elk horn, rich in essential trace minerals, is an aphrodisiac in Korea. Ranchers in Scotland and New Zealand are learning—not unsuccessfully—to raise deer for human consumption. The velvet (rich in organic iron from the animal's blood) was sold in the Eastern market as an aphrodisiac until the British Veterinary Association put an end to the inhumane shearing off of tender antlers "in velvet" of fully conscious and terrified deer. Perhaps the flavor of ground velvet appeals to people because their intuitive nutritional wisdom tells them that their bodies need such dietary supplements. (According to agricultural-nutritional analyst Peter Reich, iron deficiency is one of the most serious and widespread deficiency diseases in the United States.)

How have these trace-mineral deficiencies arisen? First, from natural erosion, which has been accelerated by deforestation, strip-mining, and intensive agriculture. And especially from accelerated extraction through cropping and not returning what we take from the land. What is returned is not blood, bones, antlers, manure, and the dead remains of animals, but artificial fertilizers—nitrogen, phosphates, and potash. This is not enough. Though they temporarily help boost yields, the crops are nutritionally deficient. These synthetic fertilizers—some of which are derived from nonrenewable fossil fuels—do not sufficiently enhance the health of the crops, so more pesticides and herbicides are needed.

Lacking organic material as well as trace minerals and

---

ancient Rome, where their bathwaters were rich in this mineral, to heal "the madness") is used today in the treatment of manic-depression.

carbon essential for humus formation, the soil—the "flesh" of the earth—does not hold moisture. The rains run off it and the winds blow away the topsoil, or it becomes quickly compacted, necessitating heavy tractor power to plow it for seeding, which further compacts the earth.

The United States holds the world record for the consumptive use of water for food production—with animal products accounting for eighty-five percent of the total.[17] A. M. Altschul calculates that the per capita per day indirect consumption of water used to irrigate crops to feed livestock and to provide water for stock is around 2,500 gallons per person per day, (a person living on a wholly vegetarian diet would indirectly utilize only 300 gallons per day).[18] George Borgstrom estimates that the contribution of livestock to water pollution is more than ten times that of people and more than three times that of industry.[19] The adverse impacts of all this, include soil degradation, desertification, groundwater depletion, and water pollution.

It has been estimated that the use of water for irrigated agriculture has tripled since 1940. Irrigation now accounts for over a fourth of the nation's crops. Groundwater supplies twenty-five percent of all water used in the United States and about forty percent of all irrigation water.

The vast underground reservoir known as the Ogallala Aquifer, underlying the Great Plains grain belt, is being depleted rapidly to irrigate fields that were meant only for dryland farming. The quality of the aquifer is now seriously threatened by fertilizers, pesticides, and salination.*

Residues from the insecticide toxaphene (see p. 68) have been found in rainwater in Bermuda and in rivers and lakes in twenty-nine states sampled nationally by the EPA.

*It would be prudent for the government to restore this region to its original prairie state, along with buffalo, antelope, and other wildlife.

(It has also been found in fish-eating birds, including the endangered brown pelican, in Louisiana.) Recently, pesticides have also been implicated in creating algal blooms in fresh water, a phenomenon that could aggravate the ecological problems of other river pollutants, particularly nitrates (from fertilizers), mercury (from fungicides, paper mills, etc.), and cadmium and lead (from sources such as automobile pollution and sewage runoff).

According to Gerald Admires, a Navy veteran of weather observation, some of the erratic weather (more frequent droughts, floods, and temperature extremes) we have recently been experiencing is the result of ongoing state and federal programs to influence the weather. He notes that there is the Winter Orographic Snowpack Augmentation Program set up between some western states and the federal government, to increase the snowpack in the Rocky Mountains and so give western states more water. This weather modification (which is done with chemicals seeded into the atmosphere) is supposed to have an effect on only a 250-mile radius, but Admires believes that the effects are far more widespread and are one factor in the increasingly severe weather in the Midwest. The United Nations World Meteorological Organization has an active World Precipitation Enhancement program. Such programs may be contributing significantly to the present instability of the earth's climate.[20]

American agriculture is greedy in its use of nonrenewable fuels. The United States consumes six times more energy per capita than the world average. Dr. V. G. Dethier estimates that the "the modern American farm uses 5 or 6 times more fuel calories than are harvested as fuel calories."[21] Fuel calories are used extensively in the manufacture of pesticides and synthetic fertilizers (since they are derived from petrochemicals, coal gas, and rock phosphate) and to a lesser degree in spreading these chemicals over the crops by tractor and airplane. It is estimated that U.S. agricultural efficiency has decreased tenfold since 1910 and uses some ten calories of energy input to produce one calorie of food.[22]

The processing and packaging of food uses as much energy as the farms that produce our food. For every $10 spent on energy to grow food, another $5 is spent moving it around. (Processing uses almost one-third of the energy used by the U.S. food system, some 1,000 calories of energy being expended for each calorie of processed foods consumed.)[23]

High energy costs are now making it less attractive to pump surface water over long distances. Where farmers rely on groundwater such as the Ogallala Aquifer, declining water tables coupled with rising pumping costs are making irrigation more expensive.

The rushing waters from mountain streams and glaciers that crush rocks and grind them into a fine powder called loess (which the wind helps further distribute across continents) enrich the soil with trace minerals. The natural loss of trace minerals not only destroys forests and jungles; it accelerates the "greenhouse effect"—the trapping of carbon dioxide in our atmosphere. These conditions are further aggravated by improper land use and the burning of fossil fuels. According to John Hamaker, this will inevitably be corrected by an expansion of polar ice, the growth of glaciers, and the cyclical appearance of a new ice age.[24]

As vegetation is destroyed, the "albedo"*—or shininess—of the earth's surface increases. This means that more sunlight and solar energy is reflected back into the atmosphere. This affects wind currents, wind convection, and rainfall patterns both locally and great distances away. Farms

*This phenomenon is also caused by large cities of glass, concrete, and steel, and by extensive roadways and urban suburban sprawl.

in Europe and North America could be affected by increased albedo in tropical forest areas subjected to extensive deforestation and experience climatic aberrations of increasing magnitude and duration, resulting in unexpected crop failures from drought, flooding, or abnormally low temperatures. Such perturbations have increased in recent years, causing significant crop losses in the United States.

The threats to public health caused by the foods we eat come primarily from two sources: chemical and bacterial contamination. A secondary problem is the unhealthiness of the standard American diet.

Hazardous chemicals used as pesticides (including herbicides) and antibiotics, growth stimulants, and other drugs in animal feeds are absorbed into the meat, eggs, and dairy products we consume. There is increased reliance upon a last-spray pesticide treatment of crops to prevent spoilage; artificial ripening with gas and petroleum, wax covering of vegetables, and greater use of food preservatives (some of which may be carcinogenic). Additional chemicals are used for their color- or flavor-enhancing and stabilizing properties.* Other compounds are employed to facilitate storage, and freezing is accomplished with other ingredient-extending and "fortifying" chemicals. (Sugar, for example, is one of the most ubiquitous of these chemicals.) The EPA is considering the ban of daminozide, a possible carcinogen that is sprayed on apples to make them ripen uniformly and so permit the grower to call in only one team of pickers.†

*Allergists have noted that many people are highly sensitive to foods that have been kept in plastic-lined containers or wrapped in plastic. Many plastics contain DEHP, a carcinogenic additive that can actually migrate from the plastic into surrounding liquid or other food material. This substance is known to cause cancer in laboratory animals and may be one of the most serious consumer-health hazards that is yet to be looked at by the Environmental Protection Agency.

†As of January 1986 the EPA decided to delay the ban of this chemical

While a few of these agrichemicals undergo rapid natural degradation, many are stored in their original form within plants (for example, dieldrin accumulates in carrot tissue) and will therefore later be ingested by us or by farm animals who will further concentrate them and then pass them on to us. Arsenic—poisonous to humans—is fed to poultry to improve their looks and stimulate their appetites. As has been noted, beef cattle are fed with poultry manure, the rendered remains of animals, and such surplus agricultural produce as oranges and pineapples, all of which already contain chemicals. Chemicals such as DDT enter into the food chain in these ways. (Some chemicals may even become more toxic when they are partially metabolized in the body: DDE and DDD are breakdown products of DDT that result in impaired eggshell formation and hatchability, especially in predatory birds.)

Countless chemicals (such as kepone, dieldrin, and DDT) accumulate in the bodies of freshwater and marine organisms, possibly lowering their viability, and it is through them, via the food chain, that we eventually intoxicate ourselves. While many of these chemicals are originally stored in body fat, under stress and during lactation they may be released into the bloodstream.

According to an extensive study conducted by the National Research Council (NRC) and released in March 1984, there are some, 3,350 pesticides in use, and most have not been adequately tested. Toxicity data were either inadequate or nonexistent for sixty-four percent of these substances. It was concluded that fifty percent of the cancer studies and seventy-five percent of the genetic toxicity experiments which had been done on pesticides were flawed and unre-

"pending further study." Daminozide is used primarily for cosmetic reasons and to control ripening and enhance storage of tomatoes, grapes, peaches, pears, cherries, prune plums, and nectarines.

liable. In essence, because of inadequate scientific testing and data, complete toxicity and health exposure assessments are available for only about ten percent of the pesticides and five percent of the food additives in commercial use in the United States.* The Environmental Protection Agency is reportedly ten years behind in its safety testing program. Much of this delay has been created by private-contract testing laboratories that falsify their reports, a fact that has necessitated the establishment of a new governmental regulatory division to ensure good laboratory practices.

According to one government study,[25] almost all poultry, 90 percent of all pigs and veal calves, and 60 percent of all cattle receive antibacterial additives in their feed, and 70 percent of all beef cattle are given growth-promoting additives (see pp. 78). The General Accounting Office has identified over 140 drugs and pesticides (of the more than 1,000 used in livestock production) that are likely to remain after slaughter as significant residues in meat.[26] Forty-two of these are known or suspect carcinogens, 20 can cause birth defects, and 6 can effect genetic damage in laboratory tests on animals. Drug residues from medicated feeds also remain in storage bins, augurs, troughs, conveyors, and other machinery, and thus can contaminate the *un*medicated feed that is supposed to be fed to animals a few days prior to slaughter. This source of contamination was believed responsible for 57 percent of the illegally high residue levels of sulfa drugs in

*Some ninety percent of fresh and frozen fruit and vegetable produce consumed in the United States between December and May is imported from other countries, especially Mexico. According to informed sources, much of this produce contains six times the level of pesticide residue considered acceptable by the USDA. Conversely, Leptophos, a highly toxic chemical whose use has been banned in the United States (see pp. 40–41), is still being manufactured for export abroad. Because it is a very stable and persistent chemical, it has accumulated in food, resulting in illness and death in humans in countries such as Egypt.

pigs tested in the last half of 1977 by federal meat inspectors.

The rise in annual production of synthetic organic chemicals since 1920 has been astronomical, from less than one billion pounds in 1920 to over three hundred billion in 1978, especially of plastics, solvents, and pesticides.[27] Some of these are known carcinogens, some suspected; few have been evaluated. Their long-term and synergistic toxicities are largely unknown. Related to this, the national incidence of cancer (excluding cancer linked with tobacco smoking and old age) has increased at an annual rate of around one percent for men and almost two percent for women. Excluding highly curable and detectable cervical and skin cancer, the American Cancer Society estimated that there would be 835,000 new cases in 1982 (an increase of 200,000 since 1971).

It has been estimated that as much as 90 percent of all human cancers are caused by environmental factors ranging from pesticides to industrial chemicals, according to the National Cancer Institute (NCI).

Heptachlor, a pesticide strongly suspected of being a "complete" carcinogen (capable of both initiating and promoting cancer), reached levels as high as 2.7 parts per million in the milk of thousands of cows in Hawaii in March 1982. The FDA's "action level"—at which the contamination level exceeds 0.3 ppm and milk cannot be sold—was clearly surpassed. Cattle had been fed chopped pineapple tops that had been heavily treated with this pesticide under EPA permit. How frequently such instances of pesticide contamination of our food chain occur is unknown. We will only know for certain generations from now, if and when the cancer epidemic in society begins to subside following drastic restrictions in the use of agrichemicals worldwide. Many, however, along with industrial chemicals such as PCB, will remain in the food chain for generations, since their biodegradation is so slow. And since cancer is such a complex disease, with a long dormancy period, accurate identification of its causal agents is virtually impossible. The problem is compounded by the fact that there are so many different chemicals contaminating the environment and the tissues of our bodies. It is rare for a specific cancer agent to be identified; one case will suffice. Well water contaminated with the banned pesticide DBCP may now endanger the lives of up to 200,000 people in California. Harmful levels of this pesticide (manufactured by Shell Oil, Chevron Oil, Occidental Petroleum, and Dow Chemical) found in people's drinking water correlated with a rate of deaths from stomach cancer twice the normal incidence.[28]

Toxaphene is an insecticide used widely to rid livestock of external parasites; it is a chlorinated hydrocarbon related to the now-outlawed compounds aldrin, dieldrin, and DDT. Some 40–100 million pounds of toxaphene are probably used annually by farmers. According to the NCI, this drug causes liver cancer in mice and, possibly, thyroid cancer in rats. Cattle have been killed with this drug from being sprayed by government veterinarians. This complex compound also contains both carcinogenic and mutagenic components. As has been noted, toxaphene has been discovered in water and fish-eating birds; it has also been found in market-basket food samples taken by the FDA; milk in Arizona; and catfish and other commercially sold fish in Louisiana.

The herbicide 2,4,5-T, widely used by the U.S. Forest Service,* and other related herbicides contain dioxins, which are highly toxic chemicals (they cause birth defects in laboratory animals at doses in the range of several hundred parts per *trillion*), and are possibly the most potent carcinogens known. Dioxin has been found in beef from contaminated

*And also by ranchers to control "weeds" on rangeland, and by power companies to control weeds along power lines.

pastures and in the milk of human mothers who live within the vicinity of dioxin-sprayed acres. Dioxin will therefore also be concentrated in cow's milk.

The accidental contamination of animal feed with PBB in Michigan in 1975 (it was mixed in place of a mineral supplement) necessitated the destruction of several thousand cattle, sheep, and hogs. PBB and PCB are used widely as insecticides and wood preservatives. Cattle that lick any treated wood in the barn or stall could ingest these substances, which contain, among other things, the highly toxic dioxins. In the Michigan catastrophe, dioxins were later found in the milk of cows that supplied the local communities.

Any chemical that is deliberately (pesticides) or accidentally (radioactive waste, lead from air pollution) broadcast onto the crops or pasture that cattle eat may be concentrated in the meat (fat) or other internal organs, or pass into the milk (especially fat-soluble chemicals in the cream) and thus be ingested by people. For instance, cow's milk has long been used as a good indicator of levels of strontium 90 fallout. Pesticides such as DDT, aldrin, dieldrin, dioxins from the herbicide 2,4,5-T, aflatoxins from moldy feeds, PCP, and PBB are concentrated in cow's milk from contaminated and moldy feed and contaminated pastures. Other drugs (such as methoprene and permetrin to control ticks and flies) used in cattle, whether administered orally or by injection, may, like the PBB's, act as immunosuppressants. Ironically, the dairy cow and the human female are emerging as the ideal "guinea pigs" for monitoring the food-chain accumulations (or "biomagnification") of such dangerous chemicals in our species.* The extent of behavioral, neurological, teratogenic, genetic,

*Several studies have shown that human breast milk is now commonly contaminated with several industrial and agricultural toxins, particularly carcinogenic pesticides such as DDT.

and carcinogenic consequences to ourselves and our offspring remain to be determined.

HCB, a seed fungicide, has been shown to accumulate rapidly in plants, molluscs, and fish (via water runoff from planted fields). This substance has a very slow biodegradation rate and causes illness in test animals, yet its use is widespread. Further, pesticide-spray solvents contain petrochemicals that may be toxic to man and have been implicated in an outbreak of Reye's syndrome in Nova Scotia children. Such solvents may interact with viruses and exacerbate disease or, like some animal insecticides, act as immunosuppressants, thereby increasing human susceptibility to diseases and "new" viruses.

Malathion is a pesticide widely used by home gardeners. It was sprayed over vast areas of California in 1981–82 to control the Mediterranean fruit fly. Yet for some years the EPA has had evidence (from two studies done in the late 1970s by the National Cancer Institute) that exposure to malathion can pose an unacceptably high risk of cancer.

According to the columnist Jack Anderson, the EPA ignored these studies, doing a scandalous "cut and paste" review of malathion and actually including only those portions of the NCI summaries that could be cited to minimize the dangers of malathion. Senior EPA officials subsequently hired the Battelle Institute, an independent research laboratory, to evaluate the EPA's review of the NCI studies. The Battelle report concluded that there was no evidence of a critical review of the NCI report by the EPA: it had been ignored. Furthermore, they said that from their analysis there was "highly suggestive evidence of a dose-related carcinogenic effect" with malathion. According to Anderson, "EPA brass stood by the review that Battelle lambasted." Meanwhile, Dr. Adrian Gross, an EPA cancer researcher, concluded from his study of NCI data that as many as one of every one thousand people exposed to "currently established tolerances of malathion" can be expected to get cancer.[29]

Larvadex is a pesticide added to poultry feed to prevent avian flu: the virus is transmitted by flies, and flies deposit their eggs in the manure piled up beneath the birds, so Larvadex in the droppings will prevent the fly eggs from hatching. The EPA gave out forty-seven emergency exemptions for poultry farmers to use Larvadex to control flies over a two-year period before it was taken off the market in August 1983 for further testing and review. Subsequently, the EPA moved to finally register this poison for all egg farmers to use as they need until December 31, 1985. Traces of a harmful chemical in Larvadex called melanine (which causes bladder stones and bladder tumors in male laboratory rats) are to be found in the eggs and meat of poultry fed this pesticide.

In late June 1984, the EPA pulled Larvadex off the market after California health officials called the agency's attention to tests showing that Larvadex is toxic (teratogenic) to fetuses even in the smallest doses given to laboratory animals. According to John Moore, head of EPA's pesticide division, "For whatever reason, the review [of these teratogenic findings] somehow didn't get done. It got lost."[30] Following this policy reversal, the EPA had to cancel several "emergency" permits allowing Florida farmers to use a similar cyromazine-based pesticide on lettuce, celery, tomatoes, and carrots.

Then, in May 1985, the EPA did a turnaround and reapproved Larvadex. Apparently the agency is satisfied with the manufacturer's latest studies on birth defects and has established residue limits of 2.5 ppm in eggs and 0.05 ppm in poultry meat. "Residue limits" are a product of legal-bureaucratic stupidity, since limits are meaningless: other chemicals present as residues may greatly enhance the toxicity of Larvadex and other "regulated" chemicals. (Ciba-Geigy also has to indicate how much of the chemical workers will be exposed to when they mix it into chicken feed.)

In November 1984, the EPA placed new restrictions on alachlor, the nation's most widely used corn and soybean herbicide, to determine if it should be banned. Laboratory studies on animals have shown it to be a potent carcinogen. The chemical, manufactured by the Monsanto Company, under the brand name Lasso, has been found contaminating streams and underground water supplies in at least three major corn-producing states. Because it is present in animal feed, there is a great likelihood that it is present at some level in meat and milk. The EPA also estimates that some 600,000 farmers and farm workers face cancer risks because of their exposure to this chemical while applying it to the fields.

EDB,* known to cause cancer in test animals, is a fumigant used to prevent grain from becoming moldy (some molds, such as aflatoxin, are carcinogenic). It was belatedly addressed by the EPA in 1984 after some state health authorities began removing contaminated cereals, cake mixes, and other cereal-based convenience foods off grocery shelves. Used extensively by fruit growers, EDB contaminates wells and drinking water in California, Florida, and other "garden" states. (Its use on citrus crops was prohibited by the EPA on September 1, 1984.)

Congressional hearings were held in June 1984 over the EPA's pesticide-regulation activities. At these hearings it was revealed that while most of the major uses of EDB were banned only recently, the EPA had been informed ten years earlier by the National Cancer Institute that laboratory an-

*EDB is also a widespread air pollutant, since it is put into gasoline as an additive.

imal studies of EDB showed it to be one of the most potent cancer-causing substances yet discovered. These hearings also revealed that the EPA has been granting more and more emergency exemptions for unregistered use of pesticides, such exemptions being used as a way to circumvent the legal prohibitions against unregistered pesticides. Furthermore, much scientific data submitted in support of current pesticide registrations have not only been inadequately reviewed by the EPA: many were actually falsified. Three officials of International Biotest Laboratories (IBL) have been indicted and convicted for falsifying data in support of several pesticide registrations that the EPA approved. It was found that only about ten percent of over two thousand IBL studies which had been submitted in support of some twelve hundred pesticide registration applications could be considered valid.

There is evidence that pesticide dangers may be increased by pollution:

> Dust and high levels of ozone (an air pollutant) increase production of paroxon from the widely used organophosphate pesticide parathion 30-fold over cleaner settings, according to Robert Spear, Yun-San Lee, John Leffingwell and David Jenkins of the University of California at Berkeley in the March–April *Journal of Agricultural and Food Chemistry*. Paroxon, formed by the addition of oxygen to parathion, is 10 to 100 times more toxic than parathion and appears to have caused at least 29 poisonings in California field-workers, Leffingwell says. Other organophosphate pesticides may react in a similar way.[31]

The World Resources Institute has recently published a report indicating that insects and other pests are becoming more resistant to chemical pesticides. This fact now poses serious public-health dangers and contributes to increased food production costs. The report notes that the number of harmful insects immune to one or more pesticides reached 428 species by 1980, almost doubling the figure of 224 in 1978. The study estimated that insect resistance now costs American farmers at least $150 million a year in crop losses and increased chemical applications. Furthermore, rodents are becoming more resistant to chemical poisons; at least 50 species of "weed" have become resistant to herbicides; and more than 150 kinds of fungi and bacteria are now resistant to agricultural chemicals, an increase from 20 in 1960.[32]

Alternatives to pesticides are not revolutionary.

> Although seldom appreciated because of the publicity given pesticides, nonchemical control methods are used more extensively to protect crops from pests than are pesticides. Nonchemical controls for insect control are employed on about 9% of the crop acres, compared with only 6% treated with insecticides. For control of plant disease, nonchemical control is used on 90% of the acreage, compared with less than 1% treated with fungicides. Mechanical weed control is used on an estimated 80% of the crop acreage, while about 17% is treated with herbicides.[33]

The problem persists that those pesticides that remain are too dangerous and too much.

The health risks of carcinogenic nitrosamines in foods such as cured ham and bacon, which form when the preservative sodium nitrite is cooked, have been hotly debated in the United States. However, the potential health risks of widely used nitrate fertilizers, which accumulate in plants and drinking water and may be converted into nitrites and nitrosamines, have been ignored. Nitrates are broken down into nitrites in our digestive systems, and these are toxic. Silage made from such overfertilized crops can cause death, abortions and other complications in cattle. When gastric hydrochloric acid is deficient (a condition known as achlorhydria), carcinogenic nitrosamines are formed from these nitrites, according to Dr. Harry Walters.[34] Babies and young animals such as piglets have little gastric acid and will develop a blood disease called methaemoglobinemia from absorbing high levels of nitrites from digested nitrates. The World Health Organization has set a recommended upper limit of 50 mg./liter of nitrate in drinking water. (Switzerland is the only country that randomly samples vegetable produce and condemns foodstuffs that contain nitrate or nitrite above permitted levels.) Farm animals given feeds high in nitrates and nitrites are known to suffer from toxemia, mineral imbalances, and vitamin A and E deficiency.

As Dr. Walters points out, a National Academy of Sciences publication in 1972 entitled *Accumulation of Nitrate* was—like many other government-sponsored efforts—very carefully phrased so as not to rock the economic or political boat. However, one of many disturbing findings in this report showed that nitrites *accumulate* in some crops, such as spinach, after harvesting. The older the produce, the higher the nitrite content.

Nitrosamines have been linked with esophageal and stomach cancer. Significantly, because of the carcinogenic risks attached to excessive use of nitrates, medicinal preparations of potassium nitrate were withdrawn from the market in France in 1981.

While deficiencies in trace elements can lead to disease (see p. 59), these minerals can also accumulate to harmful levels in the bodies of animals, as well as in certain plants. Cadmium, which in high doses causes kidney disease and other problems, has been found to reach unsafe levels in certain internal organs—the liver and kidneys especially—of farm animals. Artificial fertilizers and industrial air pollution may be one source, or leaching from the soil and contamination of drinking water. In February 1984, the USDA ruled that the kidneys of mature chickens and turkeys must be removed prior to sale. Kidneys are a permitted ingredient in mechanically deboned poultry and chicken frankfurters. Adult poultry livers were not condemned, however, in spite of evidence that cadmium also accumulates in this organ.[35] Residues in the organs of pigs and cattle aren't considered a sufficient health hazard to warrant more extensive condemnation in the United States, though concern is being expressed over this problem in Europe.[36]

High levels of zinc and cadmium in the air and soil were discovered around a zinc-smelting factory by University of Pennsylvania researchers. High levels of these elements were found in horses and foals in the region, who developed a crippling arthritislike disease. Excess zinc induced a copper deficiency in these animals, which the scientists believed was responsible for the animals' illness. Three dairy herds were also affected to a lesser degree.

In March 1982, the USDA relaxed restrictions on the amount of phosphates food processors can use in meat and poultry products to retain flavor and juices during cooking. Phosphates in canned ham, bacon, beef patties, hot dogs, bologna, and cooked sausage (they are also present in soda beverages) are a health hazard, as noted above, since excess dietary phosphate can cause osteoporosis (calcium deficiency and thinning of the bones) and other nutritional imbalances and diseases. Excessive use of phosphate fertilizers also accounts for high phosphate levels in cereals. Rock phosphate fertilizer can contain radioactive trace elements which appear in some crops, notably tobacco.

While fluoride is a well-recognized poisoner of cattle, it is added to drinking water to prevent cavities and is in most toothpastes. Recent laboratory animal tests, however, indicate that fluoride disrupts the body's DNA and weakens the immune system, which can lead to genetic damage, cancer, and allergies. It is a widespread contaminant of food, water, soil, and air, especially from plants manufacturing steel, aluminum, and fertilizer. Many pesticides and fertilizers also contain high levels of fluorides. Fluorides are especially harmful to the human fetus when the mother is not well fed: thus, fluoridation of drinking water may be especially hazardous to low-income families.

It should be emphasized that many agrichemicals, highly toxic aflatoxins (from moldy feeds) and industrial pollutants, such as PCBs, tend to be fat-soluble and therefore accumulate in animals' fat, in egg yolk, and in cream/butterfat. This problem is compounded further by the fact that animal fat is rendered in slaughterhouses and is then used in formulating poultry and livestock feeds, which can lead to further accumulations of potentially harmful chemicals in farm animals and their produce, and, via the interstate feed-distribution system, result in widespread contamination of farm animals and their consumers.

A long-standing point of concern had been the use of diethylstilbestrol (DES) and other synthetic estrogens used to promote growth in cattle. DES was injected in the animal's ear as a slow-release implant and/or was included in feed at low levels up to seven days before slaughter.

However, the use of DES as a growth promotant for cattle and sheep has now been officially banned by the FDA. This synthetic hormone, which has been shown to cause cancer in both animals and people, had been used since the 1950s to make animals grow faster. Alternatives, which are presumably safer, are now available. The agency had previously banned DES in August 1972 (in feed) and in April 1973 (as an implant),[37] but the bans were overturned in January 1974 by the U.S. Court of Appeals because the agency had not held public hearings.

Yet, in spite of this ban, over 400,000 cattle have been impounded by the government because they were given illegal DES implants. (Interestingly, if this hormone had been linked with testicular cancer rather than cancer of the cervix, it undoubtedly would have been banned long ago.)

Hormones used widely in farm animals to boost growth or induce labor and estrus (heat) for breeding are now suspect in the disturbing condition known as thelarche, or premature puberty, which afflicts some ten to fifteen thousand children (girls under the age of eight and boys under the age of nine) in the United States each year.[38]

Recently, an epidemic of thelarche has occurred in Puerto Rico. Some two to three thousand young girls have shown varying degrees of premature sexual development such as breast enlargement, development of pubic hair, menstruation, and deformed skeletal growth. A number of epidemiologists in Puerto Rico believe that this disease may be linked to the use of estrogens used by chicken farmers to fatten their poultry more quickly.[39] But there is still no definitive answer as to the cause of this disorder.

More recent studies on the abnormal sexual development of Puerto Rican children have revealed high levels of zeranol, an estrogenlike growth stimulant in the blood of these children. This growth stimulant, sold under the trade name of Ralgro, has been on the market for several years and is widely used in many countries. This hormone is implanted in pellet form under the ears of cattle and sheep, and farmers are supposed to withdraw the pellets sixty days before the animals are marketed for slaughter. According to informed sources, the government does not check to see if this is done.*

*At a meeting in Brussels in December 1985, Common Market agriculture ministers decided to ban the use of all hormones as livestock growth promoters. The ban will go into effect on January 1, 1988.

Another group of drugs given to farm animals, the nitrofurans, are known carcinogens, but attempts to get these off the market have so far failed.

Another problem with additives is that, for many of the antibiotics used in the feed, there are *no* stipulated withdrawal times (when they must not be fed before the animal is slaughtered). For example, there is no stipulated withdrawal time for neomycin, a drug that alone accounts for twenty-five percent of residues found in cattle after slaughter. Nevertheless, a group of U.S. senators, acting on behalf of the livestock and poultry industry, have introduced a resolution that would prevent the FDA from adopting strict controls over the use of antibiotics in animal feeds "pending further research."*

Meanwhile, in the United Kingdom, a Catch-22 situation seems to have arisen. While tetracyclines and penicillins are now banned as feed additives (because of concerns over the development of resistant strains of bacteria), antibiotic substitutes (such as lincomycin, avoparcin, and tylosin) are creating a new set of problems.[40] These additives, according to Dr. H. Williams Smith and Dr. J. F. Tucker of the Houghton Poultry Research Station, indirectly promote the persistence and spread of food-poisoning salmonellae by inhibiting the intestinal growth of other harmless bacteria that normally compete with or antagonize salmonellae.

Government residue test programs on animal produce do not keep unsafe meat from reaching the public. Rather, they monitor the incidence of violations and endeavor to prevent recurrences. In spite of such inadequate and costly

*In November 1985, Secretary of Health and Human Services Margaret Heckler denied an "imminent hazard" petition from the National Resources Defense Councils to ban the feeding of penicillin and tetracyclines to livestock at subtherapeutic levels.

tests, the serious problem of metabolite residues (i.e., break-down products) of drugs and other chemical contaminants in animal produce remains. These are *not* tested for, and many may be even more harmful than the chemicals that are tested for and from which they are derived.

In 1978 hearings before the House Oversight of Investigations Subcommittee of the Interstate and Foreign Commerce Committee, the FDA director, Dr. Donald Kennedy, discussed the problems of drug residues:

> Kennedy said FDA's ability to take regulatory action when an illegal residue is found in meat or poultry is severely curtailed because of:
>
> Inadequate identification of animals at the slaughterhouse;
>
> Loss of grower identity through intervening sales between the grower and the point of slaughter;
>
> Exposure to drugs and other contaminants between time of sale by producer and slaughter date;
>
> Disappearance of evidences such as contaminated feed. "The difficulties in investigating the cause of violative tissue residues have prevented us from bringing as many regulatory actions as I would like. Clearly we need to do a better job in this area," Kennedy said.
>
> Still, he said, some actions do result from investigations of tissue residues. He cited one case where feed samples collected at one grower firm indicated significant drug carryover in withdrawal feed. A followup inspection at the feed manufacturer resulted in 59 recalls of feed that was contaminated with penicillin, furazolidone, and sulfamethazine.
>
> Kennedy said that while it is important to prevent residues from occurring in the first place, it was "probably an unattainable goal" to try to remove from the market all meat and poultry containing violative residues.[41]

The General Accounting Office (GAO) conducted a two-year study of drug residues:

> The agencies involved are the Food and Drug administration, which is responsible for insuring the safety of drugs

administered to the animals and for setting limits on the amounts allowable in food; the Environmental Protection Agency, which is responsible for the safety of consumers after pesticides and toxic chemicals are used on crops and animals; and the Agriculture Department, which inspects the quality and purity of meat and poultry.

The accounting office noted that from 1974 through 1976 the Agriculture Department had reported finding illegal residues in only 2 percent of the meat and poultry it sampled. This estimate did not adequately reflect the actual situation, the G.A.O. said, contending that the figure was closer to 14 percent in all animals and as high as 16 percent in pork.

The report was more critical of the inspection procedures of the Agriculture Department and the Food and Drug Administration than of the intentions of the two agencies.

"They cannot locate or remove the residue-containing meat and poultry because the animals are sold and often consumed before sample analysis is completed," the report declared.

Furthermore, it said, the Agriculture Department did not have the authority to hold suspected meat and poultry until receiving the results of tests, and these products could not be identified once an animal had been slaughtered.

Once illegal residues are found by the Agriculture Department, the other two agencies should determine the cause, the report said.

"But during a four-year period ending in 1978, the Food and Drug Administration reported such investigations on only 37 percent of the cases referred to it by the Agriculture Department for follow up," the accounting agency said.[42]

The National Academy of Sciences reports that the present federal meat-inspection system is grossly outdated and fails to adequately monitor for chemical and bacterial contamination of meat intended for human consumption.[43] Several reasons are given for the need to improve the monitoring program. Meat has been linked with more than half of the 2,600 food-borne outbreaks of gastric illness between 1968 and 1977. In 1981, salmonella contamination alone accounted for some twenty-six percent of all reported food-related illnesses. The report also found that the USDA's program for monitoring chemical residues is seriously deficient. Far too few carcasses are sampled. In 1985, for example, of the billions of animals that will be slaughtered, the USDA plans to test for antibiotics and sulfa drugs in only 300 calves, 270 cows, 600 turkeys, and 600 hogs. Since more and more new animal drugs are being used, it makes it even more costly and time-consuming to test animal carcasses for their residues. By the time the tests are done, other carcasses on the slaughter line will have already reached consumers. It seems obvious that, until livestock and poultry are raised without drugs, the public is at grave risk from consuming farm-animal produce: the government is simply unable to control what drugs are given to the animals, and the task of finding which animals are contaminated is like looking for a needle in the haystack.

Dr. Carl Telleen, a USDA veterinarian and senior program analyst, in testimony to improve the public-health standards of meat and poultry slaughterhouse inspection, concluded that the present law is ambiguous in relation to the criteria we set for nonbacterial contamination. The standard is too low, since it is based upon an organoleptical ("look clean") standard of meat quality for human consumption, which is sufficiently high not to cause spoilage. But it is not high enough to prevent all-too-frequent outbreaks of human food poisoning—from salmonella especially, and also from E. coli. These are enteric microbes in the fecal material of animals, which are not removed in the slaughtering process.

Dr. Telleen began blowing the whistle on the USDA meat-inspection program in 1981, making public statements about the preinspection warnings and quick cleanups of slaughter plants prior to the inspector's arrival. He complained that deregulation was lowering standards, that critical reports were being changed and upgraded to make plants and local inspectors look better, and that politicians were applying pressure on inspectors to back off on their inspections.

He emphasized that, in a little-publicized USDA survey of packing plants published in 1982, some meat products were adulterated or misbranded in one of every five plants. In 32 percent of the 272 plants surveyed, it was "likely" that products failed to meet government standards even though the inspectors were giving satisfactory reports on the plants.

In letters to Senator Charles Mathias, Jr., of April 7, 1984, and Representative Tom Harkin, of June 6, 1984, Dr. Telleen explains:

> Over 200 strains of salmonella have been isolated that cause illness in both farm creatures and humans. . . . GAO in 1982 estimated that one-fourth of the federally inspected meat is contaminated. The Centers for Disease Control (CDC) estimates that over two million cases of food poisoning occur each year. The incidence of salmonella food poisoning is increasing. It leads all other causes of food poisoning. In some poultry lots, 100% of the poultry is contaminated with salmonella.
>
> Switzerland will no longer accept our exports of poultry until every lot is tested. If one lot is contaminated with salmonella, the entire lot is rejected. If two lots are contaminated the entire shipment is rejected. . . . The European Economic Community countries will no longer accept our exports if the intestines are opened and cleaned in the same room with the carcasses. But [a May 1978] regulation permits poultry carcasses with fecal material spread all over them to be cosmetically cleaned and while still *loaded with invisible* salmonella organisms to be sold as wholesome products to the unsuspecting public. . . .
>
> Industry does not trim fecal material off of product. In-

dustry employees do not sanitize hands and equipment after they are contaminated with fecal material. Neither industry or USDA test carcasses for fecal bacteria or any other bacteria. From a bacteriological, scientific standpoint, this program is the same as one that anyone could have practiced 2000 years ago. . . .

USDA still has no bacteriological sanitation program. For over 50 years, all public drinking water and Grade A milk has had to meet the bacteriological standards of quality and be tested to be sure these same fecal organisms from sewage and manure are kept out of these commodities. . . .

When salmonella food poisoning broke out in the Catonsville, Md., nursing home in December, 1983, and three people died, no one asked and no agency traced back to find out where it came from. A new bloody diarrhea E. coli disease (which is related to salmonella) broke out in 47 people who had eaten hamburgers in McDonald Restaurants in Oregon and Michigan in March and May, 1982. 41 other cases were reported to CDC from all over western and mid-western states during the rest of 1982. The cooking of the hamburger at McDonald's was blamed but *no one asked how it came into the hamburger in the first place*.

When 163 persons became ill at Circus Circus, Las Vegas on Sept. 12, 1982 after eating chicken contaminated with salmonella, they blamed the saw in the kitchen. One family in Pittsfield, Mass., suffered immediate and permanent health problems from eating salmonella-contaminated pastrami. This couple filed a million-dollar law suit against Stop and Shop Supermarkets. The meat and poultry industry owners are laughing all the way to the bank while housewives, restaurant and supermarket owners and everyone else takes their blame. USDA publishes pamphlets and sponsors children's poster contests to help consumers deal with food poisoning in the home. This is fine but *it is too bad that USDA can not sponsor a scientific inspection program so these organisms would not be brought into the home in the first place*. . . .

> The basic argument is not one of microbiological inspection per se but rather what kind of microbes is USDA trying to control. Industry is concerned with spoilage microbes and rightly so. It is spoilage microbes that reduce the shelf life of their products and turns away consumers. I am more concerned with the enteric microbes such as salmonella and E. coli in fecal material that do not spoil meat but do cause illness and death. . . . USDA refuses to make the distinction between spoilage organisms and food-poisoning organisms. . . .
>
> This brings us to the question of what kind of an inspection program USDA could or should have to control the enteric food poisoning organisms. Historically this was done by monitoring the process. We know how these bacteria come in contact with the meat and we know how to remove them—by trimming.
>
> The current controversy exists because industry wants to change this process by washing fecal material off of meat instead of trimming. I have no argument with this technology IF USDA also changes their inspection procedure. Instead of controlling contamination by USDA controlling the process, industry could control the process IF USDA inspected the end product by conducting microbiological tests. Of course these tests would have to be accompanied by tests for acceptable germicidal levels in the meat as well, to be certain industry is not substituting one health hazard for another. The problem is, that USDA has abandoned the enforcement of an acceptable process to prevent contamination but has not replaced it with a microbiological testing program of the end product.

High-pressure hosing with chlorinated water to remove fecal matter is of questionable use, since the pressured water simply drives fecal material and bacteria deeper into the tissues of the carcass. Colon bacilli also have hundreds of suction cups on their surfaces that attach to apparently clean meat, hands, and stainless-steel surfaces, and this suction, according to Dr. Telleen, cannot be broken by high-pressure hosing.

In a survey taken by *Consumer Reports* on broiler quality, one-third to one-half of the samples "gave unsavory indications of contamination with human or animal fecal matter, as evidenced by rather high counts of fecal streptococci or E. coli bacteria." No less disturbing was their finding of salmonellae in twenty percent of their samples, an alarming figure considering the serious health risk introduced by these organisms. *Consumer Reports* concluded that the USDA inspection seal is clearly "no guarantee of the bird's freedom from contamination." Concern was also expressed over the age of the flesh after slaughter. USDA considers about seven days to be the *maximum* allowable duration between slaughter and consumer purchase, but, because supermarket labels on packaged chicken indicate only a date by which the poultry must be sold (three to eleven days in the future), the consumer has no way of knowing when the bird was actually killed and packed.

Stress during transportation may cause increased shedding of salmonella organisms in pigs and poultry, and infected animals that contaminate others may well result in widespread dissemination of these bacteria at the packing plant.

Attempting to "clean" the animals prior to shipment with an antibiotic feed additive may result in potentially harmful antibiotic residues remaining in the meat. Neomycin, commonly used in Great Britain for this purpose, has been shown to have only very limited efficacy.

The Centers for Disease Control recorded almost 39,000 cases of salmonellosis in 1983, which is a 3.2 percent increase over 1982 and double the number of cases reported in 1967. The fact that more people are eating out, that the government's meat-inspection program is inadequate, and that packing plants have increased the rate of slaughtering animals (so that chances of contaminated meat entering into the food system are increased) probably all play a role in this growing health issue.

In a review of the major causes for condemnations of cattle and swine by USDA inspectors at slaughter and packing plants from 1978 to 1980, Dr. Clyde Taylor and co-workers present some disturbing figures.[44] The numbers of pigs arriving dead at the plants has increased from over 80,000 animals in 1978 to 100,000 in 1980. Death rates in calves have declined from 30,000 to approximately 15,000 and from 10,000 to 8,000 in cattle.* High incidences of pneumonia, septicemia, abcesses in pigs, and arthritis, accounted for approximately 150,000 animals being condemned annually. One to two thousand calves are also condemned annually because of injuries, and some 4,000 cattle because of emaciation. While the rates of condemnation make up a relatively small fraction of the millions of animals slaughtered each year, these figures do show that a great many animals are sick and in poor condition if not dead on arrival at the slaughter plant.

According to Dr. Barbara E. Shaw, more than half of all hogs in the United States checked at slaughter show signs of disease. One Minnesota plant found pneumonia in the lungs of 95 percent of pigs examined.[45]

As previously noted (p. 41n), the speed of inspection is too great to permit detection of all disease and defects in meat. Inspectors have only seconds to examine poultry and about one minute to check the organs and body of cattle.

Dr. P. F. McGargle, a veterinarian and government meat-inspector, has presented some disturbing speculations

*These figures are low because not all packing plants are inspected by the USDA. Plants that only sell meat locally and not to the government are exempted.

on the possibility that slaughterhouses may be contributing to human disease, notably cancer, heart disease, and gallstones.[46] Specifically, his concern is that slaughterhouse waste—excreta, blood, entrails, and the remains of cancerous and diseased animals—ends up as food for humans. A "dry melter" is used to render such wastes into a dry powder, which is then fed to pigs, poultry and calves as a protein supplement. But the temperature in the melter is set so low that many species of bacteria (and possibly cancer viruses) and antibiotic residues, as well as natural and synthetic hormones, may not be destroyed. When these are fed back to animals, they may—via this abnormal and potentially pathogenic food chain—enter the human body. In addition, cholesterol levels (which are associated with heart disease and gallstones in humans) may also be increased via this rendering-recycling procedure. Spoiled nitrate-cured meats are similarly rendered and fed back to farm livestock, a questionable practice since the nitrate–nitrosamine transformation that occurs in the human gut is potentially carcinogenic.

Dr. McGargle proposes:

> Meat processors should be compelled to return to the old, steam-pressure dilution method of rendering wastes. If this is untenable, we should at least require the temperature in the dry melter to be raised high enough to render all harmful substances inactive, and then that the material be used only as fertilizer. Present government regulations require only a temperature of 179°F for 30 minutes. This is inadequate to kill even anthrax spores, which require a temperature of 250°F for one hour.

A report by the National Academy of Sciences confirms that the use of antibiotics as feed additives can result in the development of drug-resistant bacteria that are hazardous to human health. It also asserts that such resistance could mean a serious decrease in the therapeutic effectiveness of the antibiotics that are used clinically on both humans and animals.[47]

In a letter to the *New York Times*, Professor W. K. Mass, of New York University Medical Center, wrote:

> I would like to describe our recent studies which show that there is indeed a strong link between the use of antibiotics in animal feeds and the emergence of antibiotic-resistant pathogens. Once a pathogen has become resistant to a given antibiotic, that antibiotic is of course useless for treating the disease caused by the pathogen.
>
> Our work was done in collaboration with Prof. Carlton Gyles of the Ontario Veterinary College, Guelph, Ontario. Professor Gyles has been investigating for a number of years the problem of diarrhea in piglets and calves caused by toxin-producing strains of the bacterium *E. coli*. Recently he has isolated from sick piglets that had received antibiotics in their feed 100 toxigenic strains of *E. coli*. Significantly, 90 of these strains were resistant to one or more antibiotics.
>
> It is relevant to note that the genes for toxin production and for antibiotic resistance are carried by genetic elements called plasmids, which can be transmitted from strain to strain by contact. Thus most of the 100 toxin-producing strains carried also plasmids for drug resistance. Moreover, in some strains the two kinds of plasmids had recombined to give rise to single plasmids carrying genes for both toxin production and resistance to antibiotics.
>
> The lessons from these findings are clear: (1) Many of the toxin-producing pathogenic bacteria are also resistant to antibiotics. (2) The high frequency of plasmids with genes for antibiotic resistance leads to the emergence of new, recombinant plasmids with genes for antibiotic resistance and toxin production. (3) Toxin producing *E. coli.* strains can produce disease in humans as well as animals and there is good evidence for transmission of strains from animals to humans. (4) The plasmids being considered here can be transmitted to other kinds of pathogenic intestinal bacteria and thus render them resistant.

The process by which plasmid-transferred "R-factors" result in bacteria strains that have acquired resistance to antibiotics has long been recognized. Interestingly, D. J. Fagerberg and C. L. Quarles have published a booklet *Antibiotic Feeding: Antibiotic Resistance and Alternatives*,[48] in which they suggest such alternatives as various drugs to "cure" or eliminate R-factors. Also, a strong pitch is made for the efficacy of bambermycins as feed supplements. However, the question of improving animal husbandry practices to eliminate the need for antibiotic feeding is ignored.

A study using DNA "fingerprinting" techniques showed that "plasmid molecules from human and animal isolates were often identical or nearly identical."[49] This and the patterns of human infection suggest that plasmids carrying resistance to antibiotics may be extensively shared between animal and human bacteria, and that spread of multiresistant strains of salmonella among animals and human beings may have been undetected. Health experts at the International Plasmid Conference on August 4, 1981, released a statement detailing the serious worldwide dimension of this problem, which has been found to be especially acute in the developing nations where regulations governing the use of antibiotics are virtually nonexistent.[50] The intercontinental spread of a new antibiotic-resistant gene on an epidemic plasmid has been demonstrated.[51] Therefore, we can speculate that the specter of antibiotics that are no longer useful to combat disease may well become very real.

Antibiotic residues in farm produce of animal origin may also cause allergic reactions and hypersensitivity in humans, such that penicillin or tetracycline cannot be given when needed.

Chronic recurrent yeast infections, primarily affecting women, may be triggered or perpetuated by excessive and often inappropriate use of antibiotics and steroids by physicians and farmers alike, resulting in imbalances in the bacterial flora in the human body that create the ideal conditions for such infections to develop. Yeast infections in human patients (such as with Candida albicans)—especially in the intestines, where the absorption of nutrients is disrupted and fungal toxins are taken into our bodies—have been implicated by some physicians in a variety of health problems: depression, chemical and food allergies (especially to wheat and milk), skin rashes, fatigue, anxiety, diabetes, alcoholism, and even symptoms resembling schizophrenia.[52]

More and more veterinarians throughout the United States are finding that antibiotic-resistant strains of bacteria are establishing themselves in animals on more and more farms, causing illnesses that cannot be cured by the usual antibiotic treatments. An editorial in the British *Veterinary Record* adds further weight to the validity of concern over the use of antibiotics in animal feeds:

> Risks to both animal and human health from the improper use of antibiotics have long been known. In some quarters, they have been ignored or discounted for an equally long time—in spite of the clear warnings of the Swann Committee on the use of antibiotics in animal husbandry. Now, it seems that the eventuality that Swann was concerned to prevent—the emergence of resistant strains of pathogenic microorganisms—has come about.
>
> This was suggested in a letter in our correspondence columns . . . in which G. Davies and his colleagues drew attention to the appearance of "a virulent organism causing the deaths of calves on a number of premises." It seems also that there have been cases in humans of infection by the organism—a strain of Salmonella typhimurium phage type DT *264*. The organism is resistant in vitro not only to chloramphenicol but also to sulphonamides, streptomycin and tetracyclines.[53]

And, later:

> It is two years since the epidemic spread of infection with chloramphenicol-resistant Salmonella typhimurium phage type *204* was noted in cattle and people. Since then, much has been written and said on the subject. But rather than becoming more rare, reports of human and animal infections with resistant strains of S. typhimurium have increased. The resistance patterns have shown that cross and direct selection by excessive—and sometimes illicit—use of antibiotics may lead to the rapid spread of resistance to some of our most valuable drugs. The structure of our animal industries, and human living standards and mobility, have no doubt helped an organism as robust and adept a parasite as S typhimurium to establish itself ever more firmly. The facts are all too plain, but still negligible efforts are made to control the spread of particularly damaging resistance patterns among salmonellae.[54]

B. O. Blackburn and his colleagues report that a high rate of drug resistance was observed in salmonella taken from chickens, turkeys, cattle, and swine submitted to their laboratories for serum typing during October 1981 through September 1982. Bacterial samples were tested for sensitivity to twelve antibacterial drugs. They found that three cultures were resistant to each of the drugs, and thirty percent were resistant to nine commonly used antibiotics. Multiple resistant patterns were observed for eighty percent of the bacterial cultures, with an even higher percentage in cultures from swine.[55]

Scott D. Holmberg and his colleagues have shown that antimicrobial-resistant enteric bacteria frequently arise from food animals and can cause serious infections in humans. Their investigations have shown that, in over two-thirds of U.S. outbreaks of multiple-drug-resistant salmonella infections that had a defined source, such bacteria came from food-animal populations.

> It is known that Salmonella are commonly transmitted in

> food animal products, and our analysis shows that multiple-resistant Salmonella are also frequently transmitted from animals to man. In fact, animal origins were discovered more commonly in outbreaks involving antimicrobial-resistant Salmonella than in outbreaks involving antimicrobial-sensitive strains. Thus, it appears to us that animal-to-man transmission of resistant Salmonella is not a rare event.

These investigators at the Centers for Disease Control in Atlanta, Georgia, examined 52 outbreaks of salmonella infections in humans between 1971 and 1983. In the 33 outbreaks with identified sources, food animals were the source of 69 percent of 16 resistant strains and 46 percent of 13 sensitive strains.[56] It is clear that antibiotic resistant bacteria often arise from food animal sources and represent a serious threat to human health.

These same investigators reported another study that made national headlines. They traced the illnesses of eighteen people to hamburger made from cattle fed antibiotics to boost growth. The cattle were producing large quantities of drug-resistant Salmonella Newport.[57]

A major study recently completed for the FDA by Dr. Charles Nolan in Seattle, Washington, showed that diarrheal illness is often caused by the consumption of chicken contaminated with bacteria. In Nolan's study, almost one out of every four chickens sold was contaminated by bacteria. In addition to salmonella, the bacteria Campylobacter jejeuni was implicated in human illness. Individuals who ate chicken had a twenty percent greater risk of contracting enteritis caused by C. jejeuni than those who did not. Disturbingly, thirty percent of isolates of this bacteria from animal and human sources were found to be resistant to tetracycline.[58] At around the same time that these studies were made public in the United States, over twenty-five people in England died of food poisoning, the cause being salmonella (from farm animals) that were resistant to antibiotics and contaminated the meat the victims had eaten. According to the Centers for Disease Control, two to four million cases of salmonella food poisoning occur in the United States every year.

Dr. Carl Telleen emphasizes that the poor standard of meat inspection is the open doorway through which animal fecal bacteria resistant to antibiotics can transmit their resistance to bacteria that are in our own bodies. When people get food poisoning from bacterially contaminated meat—and *this is common*—chances are high that bacterial antibiotic resistance will spread.

Dr. Stewart Levy, of Tufts University School of Medicine, writes:

> For more than 30 years in this country, subtherapeutic amounts of broad-spectrum antibiotics have been added to animal feeds to promote growth. . . . Surely the time has come to stop gambling with antibiotics. Although their use as feed additives had a major role in advancing livestock production in the past, the consequences of this practice are now too evident to overlook.[59]

Another potential consumer and farmer health hazard relates to the problem of exposure to live viruses. One strain of influenza, the swine flu virus, which first originated in Asia, now occurs in periodic epidemic waves, afflicting human populations around the world. Fortunately, humans are resistant to many of the viral diseases of farm and other animals, although for a few of these, such as rabies and psittacosis, we have no resistance. Turkeys, as well as swine, are affected by human influenza virus and can be a reservoir of infection for human beings.

Constant exposure to animal viruses and to weakened strains of live viruses (which are widely used to inoculate poultry and livestock could lead to the evolution of new strains of viruses that can cause sickness and death in humans. The close association of different animal species can mean a breakdown in species resistance. This hypothesis of the deterioration of resistance is one theory that is employed to explain parvovirus disease in dogs, which may actually be a mutant form of feline panleukopenia virus. Aujesky's disease, a terrible viral disease that causes frantic itching, self-mutilation, and death, occurs commonly in pigs, and also in cattle and sheep. It has been reported in dogs (where it is easily confused with rabies) and recently in poultry. Nonfatal human infections have resulted from laboratory accidents. In sum, the widespread use of live-virus vaccines and the high incidence of viral diseases in farm animals is a cause for serious concern. Giving animals vaccinations may also impair their immune systems and make them more susceptible to other diseases. Giving vaccines is therefore not necessarily good preventive medicine.

To cite yet another example, the milk of dairy cows can become contaminated with bovine leukemia,[60] and there may be a connection between human and bovine leukemia.[61] In some European countries, bovine leukemia must be reported to the public-health authorities, since milk from affected animals is not allowed on the market. Not so in the United States, where an estimated twenty percent of dairy cattle are infected. Though pasteurization may kill the virus, the human health risks remain to be determined. Experimentally, this virus has been shown to infect chimpanzees and human cells in tissue culture. A complex, cancerous disease called leukosis is extremely common in chickens. Some health experts now believe that since this disease is transmissible, there is a grave risk of people developing cancer from infected chicken meat and chicken franks.[62]

Irradiation of foods to kill insects and retard spoilage was approved by the World Health Organization in 1983. It is now used in twenty-eight countries, including Japan, South Africa, and some European nations.

In July 1985, the FDA approved the use of irradiation

to kill the parasite in pork that causes trichinosis in humans (freezing is now used for this purpose, although proper cooking kills the organism anyway). Previously, only a few food "ingredients" (notably spices)—not foods themselves—have been irradiated. Now, however, the FDA will probably agree to the irradiation of fruits and vegetables for insect control before the end of 1985, and the extension of the process to meats, poultry, and other foods is expected to follow within a few years. In a 1982 FDA review of 413 toxicological studies of irradiated foods, 344 (84 percent) were either inconclusive or inadequate to demonstrate safety. Of the remaining 69, adverse effects were found, according to Sidney Wolf of the Public Citizens' Health Research Group in Washington, D.C. Mice fed irradiated chicken meat have been shown to have an increase in testicular tumors, kidney disease, and cancer. Ironically, irradiated wheat, sorghum, maize, millet, potatoes, and onions produced *more* aflatoxin than nonirradiated samples after being experimentally infected with aspergillus fungus. (The EPA ranks aflatoxin as one thousand times more carcinogenic than EDB.)

A 1984 report by the Council for Agriculture, Science, and Technology supports the FDA's proposal to draw up guidelines for the irradiation of fruits, spices, vegetables and grains at doses up to 100 Krads (or 0.1 Mrads). While experts believe that doses up to 1 Mrad are safe, there have been no long-term studies done on humans.

In testimony to the FDA, Dr. John W. Gofman questioned the trend toward irradiation:

> From a lifetime of research in both heart disease and cancer, I know what sort of studies are required to ascertain the delayed effects and the cumulative effects on humans of biological agents such as diets, drugs, pollutants, and energy-transfers (ionizing and non-ionizing radiations). Animal studies and cell studies can provide information of only limited value

> about carcinogenicity, teratogenicity and mutagenicity. The kind of epidemiologic study required to find out whether or not a diet of irradiated food will increase (or possibly decrease) the frequency of cancer or genetic injuries among humans simply has not been done.
>
> What is more, such a study is unlikely ever to be done, because it would require controlling the diets of at least 200,000 humans of various age-groups for at least 30 years, and following their health-histories for at least 50 years (preferably their full lifespans).
>
> It is probable that we shall never know whether or not irradiated foods are safe. What we do know with certainty is that irradiation causes a host of unnatural and sometimes unidentified chemicals to be formed within the irradiated foods, and that the number, kind and permanence of these "foreign" chemical compounds depend on the food itself and on the dose of radiation.[63]

Research studies indicate that irradiated foods would lose much of their nutrient value and would also contain breakdown products (called unique radiolytic products) that could be hazardous to consumers.[64]

A bill before Congress at this time of writing would reclassify irradiation from its present status as an "additive" to a "process"—which, under FDA rules, requires companies to present far less information on safety and efficacy.

Ultimately, the spread of irradiation will probably depend more on economic than on health concerns. Estimates indicate that it will add 1.5 cents to the wholesale cost of a product, but this may be countered by reduced spoilage and longer shelf life.

The largest question will be consumer acceptance. Irradiated foods sold at the retail level must now bear the label "treated with gamma radiation,"* but new FDA regulations

*Spices, as a food "ingredient," do not need to be labeled that they are irradiated.

will be promulgated in the fall of 1985.* Martin A. Welt, president of Radiation Technology, fearing that the word "radiation" would encounter strong public resistance, proposes instead a label of a circular symbol enclosing rose petals![65]

*A study by the House Government Operations Subcommittee on Intragovernmental Relations and Human Resources entitled "Human Food Safety and the Regulation of Animal Drugs" (January 1986) documents the FDA's inadequate monitoring of the use of toxic drugs and nutritional supplements in raising livestock and poultry, which poses a grave threat to the health of consumers, since many of these substances have been identified as causing cancer and other illnesses and are found in beef, pork, poultry, eggs, and milk. Thousands of drugs and feed supplements in wide use have never been approved by the FDA. The agency was also faulted for failing to restrict or ban several animal dsrugs that are carcinogenic and for having identified and inventoried only 7 percent of the thousands of animal drugs now on the market, which is a direct violation of a 14-year-old statute mandating this agency to maintain an inventory of all approved animal drugs.

According to the report, 90 percent of the 20,000 or more new animal drugs on the market have not been approved as safe and effective. One drug, the antibiotic chloramphenicol, has been prohibited for use on food-producing animals because it can cause an incurable anemia in humans. Yet this drug is still being widely used on cattle and hogs. The nitrofurans, a widely used group of drugs in farm animals, have long been known to be potentially carcinogenic. The proposal to withdraw these drugs was made more than 14 years ago, but hearings were scheduled by the FDA in 1985.

4

The Nutritious Diet

THE average person in the United States consumes approximately 600 pounds of meat, eggs, and dairy products per year. This provides 70 grams of animal protein per person per day in addition to the 32 grams of plant protein consumed. If consumption of animal proteins were cut in half, total protein intake would be 67 grams—still well in excess of the 56 grams suggested for a 155-pound man by the 1980 RDA (recommended daily allowance).[1] Total world food production is 5 pounds per person per day; a human being needs only 1.3 pounds per day.

Domestic animals eat enough food for two to eight billion people. Twenty to forty percent of all food produced is lost (to pests, spoilage, etc.) and some two billion people eat enough to sustain three to four billion. Of the world's production of consumables in 1977, 737.48 million metric tons of sugar cane were produced, which is of little nutritional value. The second most highly produced commodity was cow's milk (409.09 million metric tons), followed by wheat (386.60), rice (366.51), soybeans (77.50), fish (73.50), beef and veal (46.25), pork (43.81), eggs (25.08), and poultry (24.39).[2]

Americans eat eight to twenty-eight percent of all meals outside the home, and we are eating more at fast-food outlets, according to a 1982 survey by the USDA. People are spending nearly one-third of their food dollars at such restaurants. Their menus do not provide the recommended daily allowances of vitamins, carbohydrates, calcium, and iron. The high salt, sugar, fat and refined flour content of fast foods are not in the best interests of consumer health.

The nutritional quality of agribusiness-produced crops may be far below what the consumer anticipates. In a comparison of vegetables grown by agribusiness with those grown with farmyard-manure compost, crops from the latter were higher in total sugar, vitamin C, and protein. It is worth noting that the high use of nitrate fertilizers increases the water intake of crops, which serves to reduce the nutrient content relative to weight and also decreases the storage capacity. And, as pointed out earlier, artificial fertilizers can disrupt the trace-mineral balance of the soil, which reduces the nutrient value of crops necessary for farm-animal health and consumer health alike.

Francis Chaboussou, a French agronomist, has advanced the theory of trophobiosis,[3] which suggests that present agricultural practices create health problems in plants somewhat analogous to the way in which I have shown that husbandry practices create serious health problems in intensively raised farm animals.[4] Chaboussou's theory of trophobiosis is an explanation of why pests attack crops. These pests can be insects, nematodes, and diseases, which are controlled with various poisons. According to his theory, these pests can only survive on plants that have an excessive level of amino acids in their sap or tissues. This excess can be caused by inhibition of protein synthesis, the predominance of proteolysis over protein synthesis, or the excessive production of amino acids. These internal metabolic imbalances in crops are caused by pesticides or by imbalanced nutrition in the soil, where an oversupply of nitrogen from soluble fertilizers leads to excessive production of amino acids in the plants themselves. According to Chaboussou, proper nutrition for the plants and cutting back on the application of pesticides would cause the plants themselves to be much healthier and would enhance their resistance to diseases and pests, making the need for chemical control something of the past. And, with proper soil nutriton, the nutritive value of crops would be enhanced, which is in the consumers' best interest.

Little research has been conducted to compare the actual nutrient values and quality of intensively raised farm-animal products with conventionally raised "free-range" ones. Two studies, however, *do* reinforce the concerns of many conscientious consumers that factory-farm products may be inferior and that one does not necessarily get one's money's worth or the quality produce that one expects. Freshness is an important factor for many dairy and egg products, since proteins, vitamins, and other essential nutrients may deteriorate with time; indeed, they may already be partially degraded from certain processing procedures. R. J. Williams and his colleagues maintain that the full (trophic) value of food

> cannot be ascertained from food composition tables because only a smattering of the necessary information is furnished. A food cannot support life if it is missing, or deficient with respect to, any one of the necessary nutrients. A tabulation which includes only a few nutrients—e.g., calcium, thiamin, riboflavin, niacin, phosphorus and iron—can be woefully misleading, especially if these individual nutrients have been added by way of fortification.[5]

They go on to demonstrate that barnyard eggs are superior to battery eggs, but note that whether the eggs are fertile or not makes no difference in nutritional quality.

A. Tolan and his fellow researchers in Great Britain report that they also found differences in the quality of eggs produced under battery, deep-litter, and free-range conditions. While many of the nutrients analyzed showed no significant differences according to housing conditions, free-range eggs did contain more vitamin B_{12} than deep-litter or battery eggs and fifty percent more folic acid than battery eggs. The authors conclude that these nutritional deficiencies could have significance for people who may depend upon eggs as an important source of these nutrients. This study also showed that there can be great differences in nutrient quality among eggs obtained from different farms, a finding that reflects the quality of the feed given to the laying hens.[6]

Beef consumption in the United States leaped from 58 pounds per person in 1940 to 116 pounds in 1972. In 1976, we were eating 94 pounds of beef and 52 pounds of poultry, but by 1984 beef consumption had fallen 17 percent while poultry consumption climbed 29 percent.

Dr. Mark Hegsted, administrator of the USDA's Human Nutrition Center, points out that there is no evidence that people are healthier if they eat more meat; in fact, eating *less* meat may be more conducive to health. Animal protein (meat) elevates serum lipid levels and can thus contribute to higher cholesterol concentrations and arteriosclerosis. Epidemiological data also link meat consumption with breast and colon cancer; high-protein or high-phosphate diets (phosphates are used in processed meats and soft drinks) contribute to osteoporosis (thinning of the bones)* in elderly people

*Professor Eldon W. Kienholz of the Department of Animal Science, Colorado State University, Fort Collins, is researching his plausible hypothesis that levels of the active form of vitamin D (DHD) are high in meat from stressed animals and may be responsible for the widespread consumer incidence of osteoporosis (personal communication).

and in women, especially those between forty and forty-five years of age. Hegsted notes that Seventh Day Adventists, who eat no meat (nor do they smoke or drink) but who do eat eggs and dairy products, show only half the incidence of cancer and a lower incidence of heart disease than the rest of the populace. Mormons, on the other hand, who also refrain from smoking and drinking but who do eat meat, have a higher incidence of chronic diseases than Seventh Day Adventists.[7]

Research at the Harvard University School of Medicine studied the effects of giving very lean meat to a group of vegetarians. After a few days, their blood-pressure readings and cholesterol levels increased significantly, but these values returned to healthy levels a few days after they had resumed their vegetarian diets.

A report released in May 1984 by the American Heart Association Journal on dietary changes to help lower blood cholesterol and to decrease the incidence of arteriosclerosis and heart attacks, firmly recommends we eat less animal protein, especially red meat (beef, lamb, mutton, pork, and ham), and less animal and dairy fat. Instead, we should eat poultry, fish, meatless main dishes, and plenty of fruits and vegetables.[8]

Dr. Sherwood Gorback and colleagues at the New England Medical Center, Tufts University, have found that a high-fat diet (both animal and vegetable fats) correlates with a high incidence of breast cancer. A reduction in fat intake combined with eating high-fiber foods will help protect women from this widespread disease. Vegetarian women excrete more estrogens in their urine than nonvegetarians, which might be a beneficial dietary consequence.[9]

Dr. J. A. Scharffenberger believes that a reduction in consumption of animal products could lower the incidence of heart disease by as much as eighty-eight percent and the incidence of cancer by as much as fifty percent (the three major dietary factors in cancer causation are obesity, animal fat, and lack of fiber or whole grains).[10]

Apart from the omnipresent hazard of drug and chemical residues in farm-animal produce, there are some other factors that lower the nutrient value of such produce to the extent that consumers aren't really getting what they pay for or think they are buying. "Finishing" beef cattle on grain can produce a carcass of thirty percent fat and fifty percent lean; such high-energy feeds lower nutritional quality by increasing the amount of storage-type fats.

Processed foods may be euphemistically labeled, or may contain materials we would not wish to eat if we knew about them. The American Farm Bureau Federation takes this stand:

> We oppose regulations requiring that individual ingredients of less than 2 percent of the product be listed by percentage in descending order of amount as well as those regulations requiring percentage of the total product listings.
>
> We further recommend that powdered bone, which is mostly calcium, be listed as calcium in its order of prominence in the ingredients statement (in mechanically processed meat products).[11]

However, bone meal can be heavily contaminated with lead, which accumulates in the bones of animals fed crops contaminated (via rain) with lead from automobiles burning leaded gasoline.

Milk comes mainly from Holstein cows, which secrete a milk that is high in water and low in "solids not fat"—protein and carotenes. These latter are much higher in Guernsey and Jersey cows. The differences are in part genetic and in part related to the kind of diet Holstein cows are fed to produce a low-protein (i.e., low-nutrition) milk that is high in butterfat (hence the dairy surpluses of butter and cheese).

Homogenization, which thickens the watery milk and gives it the illusion of richness, has been found to be a serious, nationwide consumer health hazard. Dr. Kurt Oster and Donald Ross have evidence that homogenized milk contributes to cardiovascular disease, especially clogged arteries, and is probably a more widespread, long-term threat to consumer health in the United States than cigarette smoking.[12]

While dairy products are a good source of protein, Dr. F. A. Oski and J. D. Bell point out that drinking milk may not be good for many people, especially children.[13] Milk is a much-touted source of calcium "for strong bones and teeth," but there are many potent alternative sources—almonds, collard greens, broccoli, and kelp. Oski and Bell's concerns about milk (aside from the tremendous subsidies taxpayers provide the dairy industry) include lactose intolerance: raw milk is not digested by many people, especially non-Europeans. Milk can also cause a tension-fatigue syndrome (often with a "stuffy nose") in children, and insomnia, anxiety, or depression in adults who are allergic to it. It has only recently been recognized that food allergies can be manifested as changes in personality, emotions, and general sense of well-being. The most common offenders are milk, corn, cane sugar, wheat, and eggs.

As for why a warm glass of milk helps one get to sleep at night, cow's milk, like human milk, contains natural opiates—a pleasant "fix" for calves and babies too, but as Oski and Bell conclude, "Cow milk is for calves." The problem of "not getting enough protein" (especially for children) can be easily overcome by eliminating some farm animal products in favor of vegetable protein from beans (including soybean products such as tofu or bean curd), lentils, nuts and peanut butter, avocadoes, and similar produce that have, ideally, been grown organically.

There has been much concern expressed over the possibility that fats in eggs and meat cause high blood-cholesterol levels. However, some nutritionists now believe that high sugar and alcohol intake are the true culprits. (It is still a good idea to remove as much fat as possible from poultry and other meats prior to cooking, since pesticides and other fat-soluble agrichemicals accumulate in fatty tissue, as well as in the liver, especially of chickens.)

The proliferation of yeast-related problems outlined on p. 90 has been associated with foods containing sugar, yeast, and vinegar. Also implicated are alcoholic beverages, malt products, dried fruits, cheese, sour-milk products, and smoked and processed meats that have mold or fungal contaminants. Improper diet and the antinutrient refined sugar (high in "junk foods") have even been linked with violent and criminal behavior. A high-fiber diet—especially of whole cereal grains, rice, oats, barley, wheat bran, and whole-wheat bread—and good-quality pasta (which is not fattening), along with legumes (beans, lentils) and fresh vegetables, all grown organically, comprises a diet that is conducive to good health.

Probably the most serious aspect of our dietary habits is our excessive consumption of wheat-based foods along with the "four white poisons"—saturated fats, salt, sugar, and refined flour. For many people, dairy products may be a fifth "poison." Some nutritionists emphasize that our diet should be more like our ancestral gatherer-hunters: a little bit of everything and not too much of any one thing. The most natural diet for people of European origin is probably one that includes far more root crops and tubers* from soils rich in trace minerals (meat lacks many of these, especially when raised on crops and forage from deficient fields): some high-energy polyunsaturated fats from nuts and seeds in winter (domestic-animal fats are now more saturated than the fat of

*Some of these, notably the carrot, actually concentrate pesticides in their roots from the soil if not raised organicallly.

wild animals). Aside from the intuitive sensibility of eating fruits and vegetables only in season (and that way one also avoids chemical poisoning from imported fresh fruits and vegetables), it may well be the saving grace of our civilization, nutritionally, to supplement (but not oversupplement) our year-round diets with high-protein vegetables (nuts, pulses, tofu). The health hazard of aflatoxins in nuts, peanut butter, and contaminated grain (which is often fed to livestock and put into pet foods) is now being recognized by the government. Better handling of produce to prevent spoilage and the proliferation of aflatoxin-producing mold, along with routine testing for this highly carcinogenic toxin, especially in milk, have been long overdue.

The unacceptable way of raising veal: young veal calves are chained to narrow crates. They quickly grow too large to turn around or make other normal postural adjustments. (Courtesy of Gail Ann Eisnitz, The Humane Society of the United States [HSUS])

This is a typical beef feedlot operation, with large numbers of animals kept in open pens lacking both shade and shelter. (M.W. Fox/HSUS)

These pigs are kept in total confinement on concrete floors without bedding. There is extremely limited

A sow is maintained in a farrowing crate on concrete with no opportunity to make normal postural adjustments or interact with her own piglets. This system is especially cruel when sows are crated for longer than a few days and subsequently kept in gestation stalls. (M.W. Fox/HSUS)

Battery cages come in many designs; some, as on this large West Coast egg ranch, are poorly designed and not conducive to the birds' overall welfare. (M.W. Fox/HSUS)

Laying hens in overcrowded battery cages endure extreme confinement typical of most battery-cage layer operations. (M.W. Fox/HSUS)

Fully conscious broiler chickens are inhumanely shackled on a conveyor en route to electrical stunning and slaughtering. (M.W. Fox/HSUS)

Industrial monocultures, San Joaquin, CA., where agrochemicals create poisoned "green deserts." (M. W. Fox/HSUS)

Wasteful irrigation using finite aquifer reserves to raise livestock feed. Colorado. (M. W. Fox/HSUS)

5

The Matter of Conscience

THE manner in which animals are being raised on factory farms has been described in Chapter 1: lifetime confinement in high-tech buildings, where they are fed artificial diets crammed with drugs. The high mortality rates on the farm and the huge number of animals that arrive dead at the slaughterhouse evidence the physical suffering to which the animals are subjected. Their emotional stress is apparent in the measures that must be taken to counteract the symptoms: debeaking, dehorning, drugs to combat increased disease susceptibility, and so forth.

The welfare of farm animals is a major issue in the public eye in the United States, Canada, Australia, and Europe. (ABC's "20/20" feature of 1980 on animal rights received more mail than any other program they have done on any issue.) Agribusiness publications have proclaimed farm-animal welfare as "*the* issue of the 1980s—and it isn't going to go away." Public concern resulted in the Mottl bill (H.R. Res. 305) being introduced in Congress, legislation that would establish a Farm Animal Welfare Committee to objectively review major concerns, gather research data, and encourage further applied research where needed. The bill was defeated.

There are several philosophies concerning animal rights, and these need to be kept distinct: (1) the animal-liberation/vegetarian movement; (2) the animal-rights/animal-welfare movement (which includes vegetarians and nonvegetarians); and (3) the fundamentalist, Cartesian, or dominionist opposition to the first two.

The animal liberation vegetarian movement sees humane husbandry systems as a compromise. For them, farm-animal welfare is irrelevant, because they believe that farm animals have a *right* not to be eaten. To T. Regan, an abolitionist, there are no morally relevant grounds for treating animals in any different way from the ways that we treat our own kind. He establishes a very convincing case that, while animals are not moral *agents*, they are moral *objects*, possessing inherent value or intrinsic worth (which is held to be quite distinct from any extrinsic value they may have to us), and are thus worthy of our moral concern and are entitled to rights. But he does not go so far as to articulate exactly what rights they should be accorded.

With reference to vegetarianism, "The question of the obligatoriness of vegetarianism . . . can arise only if and when the animals we eat are the kind of beings who have interests."[1] His critique of the limitations of such concepts as kindness and cruelty, which are widely used by animal welfarists, is extremely perceptive and reveals how patronizing and human-centered these concepts are.

Regan makes a very convincing argument for animals being accorded legal "personhood" and thus legal rights, since other nonhuman entities such as corporations and ships have been given such legal standing. He concludes by observing that

> just as blacks do not exist for whites, or women for men, so animals do not exist for us. They are not part of the generous accommodation supplied by a benevolent deity or an ever-so-thoughtful Nature. *They* have a life and a value of their own. A morality that fails to incorporate this truth is empty. A legal system that excludes it is blind.[2]

Dr. Peter Singer[3] and other philosophers use degree of sentience as a criterion to draw a line between what one should and should not eat. In fact, this guideline may be a form of overcivilized ignorance, or at least an anthropocentric/anthropomorphic valuation. If the animals have been humanely raised, transported, and slaughtered, or are "actualized" wild animals that were killed instantly—i.e., there was no suffering prior to our eating—then surely the sentience argument breaks down. If, however, one argues that the more sentient have more of a right to live than the less sentient, then we are back to an anthropomorphic projection and anthropocentric valuation.

The right not to be eaten (and other animal rights) should be cast in an ecological (but not necessarily purely utilitarian) framework, which includes natural environmental, demographic, socioeconomic, and cultural variables. Without such a framework, rights become absolute rather than relative, and could lead to disastrous ecological and social disturbances.

The cycles, densities, and distributions of coadapted plant and animal species (and the presentient substrata of earth, air, and water), within the totality of interdependent biofields that constitute the ecosystems of earth, are where the hierarchy of relative rights should be formulated. To separate sentient animals from nonsentient plants and soil microorganisms is like separating warm-blooded from cold-blooded animals, or humans from other nonverbalizing animals. This drawing of arbitrary lines is philosophically, biologically, and ecologically incompetent and misleading. All "lines" in the time-space continuum of evolution and ecology are interconnected, and to attempt to partition off plants from animals, animals from people, or blacks from whites on the basis of sentience, intelligence, color, or whatever is untenable. Phyletic, "speciesist," and racist divisions are all tarred with the same narrow enculturated view of anthropocentrism, no matter how humane and ethical the legal or philosophic intentions may be. A more nondiscriminatory, nondualistic, and integrative orientation is needed—one that is transpersonal and transpecies, and that embraces a nondiscriminatory reverence for all life (both sentient and presentient). It is only within the total framework of eco-ethics that the significance, value, and rights of each and every individual, community, and species can be apprehended.[4]

It is reasoned by some opponents of vegetarianism that, since farm animals are part of the ecology of food production, their elimination (as favored by certain animal rightists and ecologists) would seriously disturb some ecosystems. Yet these ecosystems—like the open ranges of Texas and Arizona, and the moorlands of Scotland—are unnaturally impacted by the presence of too many cattle and sheep. They would actually be more productive if the livestock were replaced by indigenous wildlife.

This group also points out that eighty-five percent of the world's draft power for agriculture comes from "beasts of burden." Would the liberation of those animals mean mass starvation amid plenty? For some farmers, tractors are not always economically or ecologically viable alternatives.

Further, by not stewarding a diversity of plant or animal species, we could cause irreparable ecological changes. Favoring less sentient (or less evolved) species over the more sentient (for human use) could lead to a world comprised only of people and plants, since, without population control, we will continue to compete with other animals for food and land.

Vegetarianism may, in many cases, be a reaction against the idea of mankind acting as an "apex predator." Meat eaters are often denigrated as "carnivores," yet the natural world cannot support us except at the expense of wildlife displaced by those farm animals that are raised for human consumption. Vegetarian or omnivore, until we can regulate our own population explosion and develop ecologically sound, regenerative agricultural practices, the more we shape the land to feed ourselves, the more we will reduce the diversity of life on earth.

At the opposite end of the scale are those who hold that we have an absolute right to exploit animals.

It is currently acceptable to assert that animals, at least mammals, are structurally, functionally, and even behaviorally similar to man—but not emotionally, for to imply that there might be any similarity in emotional experience is generally judged to be anthropomorphizing. This Cartesian attitude sees animals as unfeeling machines. Acceptance of the former assertions is considered to be based upon factual, material evidence. However, rejection of any emotional similarity is based upon two flawed beliefs. The first is that only that which is physically manifest is real, or can exist. Emotions, being nonmaterial, cannot be objectively quantified, thus they are condemned to nonexistence. But discounting of the emotional realm of subjective experience limits our human capacities for empathy, compassion, and even, paradoxically, objectivity. Second, this rejection is based upon the belief that human animals are superior to other animals, which is why we feel that we are justified in exploiting animals as we choose. Considering ourselves superior, and therefore qualitatively different, it is unthinkable—if not threatening—to propose that the emotional, subjective world of animals may be, in many respects, similar to our own.

Related to these beliefs is the recent surge of interest in creationism and opposition to evolutionary theory. Some people believe that humans are a superior creation in respect to other animals and so reject the evolutionist's view of a continuum among creatures, and therefore the idea of biological and psychological affinity between human and nonhuman animals. This kind of belief system may help to assuage guilt over the exploitation of animals.

The animal-welfare movement has been criticized as "a cult bent on destroying certain Judeo-Christian concepts," to quote Tom Simpson, representing the Parity Foundation of Detroit, a "nonprofit nonpartisan foundation committed to the interests of independent owner-operators" (animal dealers, etc.). The animal welfarists' concern over the physical and emotional well-being of animals is conceived as "part of a larger organized attack on the Judeo-Christian concept that man has unequivocal rights over animals."[5]

Others in society hold that animals and nature are ours to exploit as we wish, regardless of ethics or reason, a contention thinly veiled by claims that it is morally wrong to deliberately harm animals because such wrongdoers are more likely to be bad citizens, and that reason alone will suffice to prevent us from destroying the environment. Such moralizing is backed by the *hubris* of interpreting the term "dominion" from the book of Genesis as a free license to exploit all of creation, rather than as an injunction to responsible stewardship. Several people who hold high positions in public office today—such as Pope John Paul II—have expressed such beliefs in man's dominion (as exploitation), i.e., that nature is simply another resource for human consumption, and that animals were created for man's use.

Thus animals cannot conceivably have rights or interests and be worthy of moral concern and equal and fair consideration.

Ironically, while many scientists adhere to the theory of evolution, they label as "anthropomorphic" any concerns about the suffering and emotional well-being of animals. Yet evolutionary theory embraces the likelihood that there has been an evolutionary continuity of mental experience.[6] Thus, it may be fair to deduce that such scientists and others of like mind condemn unnecessary animal suffering as "morally wrong" only when they can find no overriding utilitarian justification. This is a weak argument. A strong one, which they cannot accept, is that because of our physiological and emotional similarity to other species (as is shown morphophysiologically, and as is implied in the theory of evolution), we should not treat them in ways that we would not treat our own kind and cause similar or greater pain or emotional distress, unless it is absolutely vital for us to preserve our own well-being or survival (rather than merely to satisfy our curiosity, or our "quest for knowledge for knowledge's sake," or for purely economic or recreational purposes).

A united front is clearly needed to oppose such flawed anthropocentric beliefs, and indeed animal-welfare, conservation, environmental, and various "rights" groups are beginning to coalesce. Some scientists may continue to remain in the middle, not to protect their impartiality and peer credibility, but rather because they are neo-Cartesian creationists and fundamentalists dressed up in Darwinian clothes, giving mere lip service to evolutionary theory or using it to prove our "superiority" over animals and our right to exploit them as we choose.

As we have shown, some members of the animal-rights/animal-welfare movement believe that animals are an essential part of our diet and ecosystem. More and more, however, are being forced to the conclusion that, as matters stand now, it is impossible to raise, transport, and slaughter livestock humanely. But even those who have reached this conclusion and have adopted vegetarianism as a personal, humane alternative (one not based on the flawed arguments we have discussed earlier) recognize that vegetarianism is not for everyone and that its proponents will remain a minority for a long time to come. The urgent task, therefore, is to seek more humane ways of supplying an omnivorous society, even while bearing in mind that completely humane meat-eating is a contradiction in terms.

For instance, proponents of intensive confinement—while condemning anthropomorphism in their critics—contend that animals, like people, are happier in a controlled environment.

> When the August sun beats down on bare lots and uninsulated steel roofs, automatic fans and sprinklers go on and the environmentally controlled beasts have their day of comfort while the Four Seasons Group pant in dusty wallows. . . .
>
> [In winter,] they huddle in groups or piles until forced by hunger to scatter with humped backs to forage or grab hasty mouthfuls from wind-swept feeders. They mill in impatient mobs while the ice thaws from a tank with a failing heater, and tiny feet leave hoof marks in the slush around the waterers. At such times you can easily imagine any of them signing away his birthright for a chance at those warm slats and nipple waterers.[7]

In other words, concludes this writer, the animals' "only problem is boredom."

Some farm-animal welfare advocates oppose intensive confinement systems and see the raising of livestock in open fields and rangeland as a humane alternative. But there are several concerns that need to be addressed. Aside from animals being exposed to climatic extremes, predation, and parasites, they generally receive less human attention and thus veterinary care when needed. Suffering and death rates are often high when these animals give birth and are denied proper care or are subjected to severe weather conditions. Starvation is not uncommon during periods of drought, a problem compounded by overgrazing. These animals are subjected to the stresses of transportation and of being rounded up for such routine procedures as branding, castration, vaccination, and dipping. These welfare concerns have been the subject of a recent report by the Australian Veterinary Association.[8]

Farm-animal welfare does not mean that animals must be put back into mellow meadows and pastures green. There are other alternatives. Husbandry systems (such as open-front climatic houses and semiconfinement units) need to be researched and more efficient means of food production (as by developing hydroponics and high-protein algal cultures) developed, since not all "natural" ecosystems can provide sufficient foodstuffs to support the world population.

Belief in the biological adaptability of farm animals to intensive-husbandry conditions has no scientific, empirical basis. This is because the major measures of adaptability and "fitness"—growth, productivity, and health—are confounded by drug, genetic, surgical, and environmental influences. Such manipulations make studies on the adaptability of farm animals under factory conditions extremely difficult. To accept that our treatments actually enhance their adaptability completely overlooks the fact that without such treatments the animals could not "adapt" and would succumb to disease because their environments would take them beyond the limits of their own innate adaptive capacity. In other words, if farm animals were biologically capable of adapting to intensive-husbandry systems, then such treatments would be unnecessary.[9] The absence of obvious clinical disease is no indicator of adaptability or well-being.

We could probably eliminate much suffering—natural and imposed—by using mind-altering drugs, psychosurgery, and even genetic engineering. But would pigs still be pigs, and chickens, chickens? The degree to which an animal is deprived of developing, expressing, and experiencing its own nature is a major welfare concern. In other words, if a domesticated chicken or pig even has half a mind left to be a chicken or a pig, should we not allow it to express some of its "pigness" or "chickenness"? How far can we go, ethically, in turning an animal into a production biomachine and thus denying it the opportunity to experience its own existence?

Animals have an intrinsic nature and interests (needs, wants, etc.) of their own, and have intrinsic worth independent of the extrinsic values we may project or impose upon them. These interests may be construed as their rights or entitlement.

Their physical, emotional, and social needs constitute their intrinsic nature, or "animalness" (which has an evolutionary and genetic basis), which qualifies them for just treatment and moral concern. No husbandry system should deny the environmental requirements of the animal's basic behavioral needs. These rights or needs should include the following minimal environmental requirements:

> Freedom to perform natural physical movement;
>
> association with other animals, where appropriate, of their own kind;
>
> facilities for comfort-activities, e.g., rest, sleep, and body care;
>
> provision of food and water to maintain full health;
>
> ability to perform daily routines of natural activities;
>
> opportunity for the activities of exploration and play, especially for young animals;
>
> satisfaction of minimal spatial and territorial requirements, including a visual field of "personal" space.

> Deviations from these principles should be avoided as far as possible, but where such deviations are absolutely unavoidable, efforts should be made to compensate the animal environmentally.[10]

Certain husbandry practices are morally repugnant to the sensibilities of most consumers, but at the grocery store we have no way of knowing which meat, eggs, or dairy products come from humanely treated animals.

There is an old American Indian tradition that the hunter must kill the deer swiftly, because if it dies slowly and suffers, he will feed its fear to his family. Christians say a traditional grace, giving thanks for the food before it is eaten; yet much of the farm-animal produce we consume is tainted with suffering and fear. No Christian can say grace, in good conscience, once the facts are known about how animals are kept on factory farms today. Jewish law requires that an animal be whole and healthy when it is slaughtered. Orthodox Jews will be shocked to learn that most of the kosher meat they buy is not actually kosher, for the animals have suffered acutely and were injured (by being shackled and hoisted by one leg) just before they were killed, which is prohibited in the Talmud.

The major welfare issues that producers of livestock, poultry, and eggs need to address are:

Overall: Problems associated with crowding or overstocking; inadequate provision of first aid and veterinary care for sick and injured animals and for humane destruction; poor preventive-medicine and preconditioning programs, and lack of veterinary treatment on badly run farms; inadequate provision for animals' behavioral requirements in many confinement systems.

Swine: Transportation, handling, slaughter, and overstocking; ventilation and floor surface of finishing hogs; continuous tethering or single-stall penning of sows and gilts; design of farrowing crates and battery cages; tail docking and care of early-weaned piglets; inadequate light for inspection.

Veal calves: Raising of calves in separate stalls or crates on slats, deprived of normal movement, iron, and roughage; feeding only twice a day. This present (European-developed) system for raising milk-fed veal is unacceptable.

Sheep: Impact of predators, notably coyotes and dogs; indiscriminate predator control; certain sheep-confinement systems and feedlot-style finishing operations.

Dairy cows: Western feedlot-style dairies (lacking shade, shelter, and lying areas); prolonged tying to stanchions; rearing of replacement calves in single, narrow pens without bedding; dehorning and castrating without anesthetic; poor handling and care of sick and injured cows and calves en route to slaughter.

Beef cattle: Lack of shade, shelter, dry lying areas, and rubbing posts in feedlots; lack of roughage in diet; transportation of young stock from southeastern states; care in pens at auctions; castrating and dehorning without anesthetic; hot-iron branding; total confinement on slatted floors.

Broiler chickens (and turkeys): Ventilation and climate control of confinement buildings; catching, transportation, and slaughter; overstocking.

Laying hens: Design of most present battery-cage systems; overstocking; debeaking; methods of forced molting; transportation and slaughter; methods of destroying culled chicks at hatcheries.

Farmers have generally been presented the issue of farm animals' welfare under intensive confinement by the agricultural press in an overly emotional and sensational way, thereby conveying the wrong impression that humanitarians believe them to be deliberately cruel or indifferent to their animals and concerned only about making money. With few exceptions (such as *Acres USA* and *Drover's Journal*), agribusiness publications have told farmers that the animal-welfare movement is going to cause them economic ruin if they do not oppose it and support their producer associations (which are tied in closely with the interests of agribusiness corporations—notably feed, drug, and equipment manufacturers). Defense of the status quo is understandable in agribusiness publications, since articles promoting animal welfare could mean a loss of advertising revenues from those companies whose interests lie in maintaining and expanding intensive-confinement husbandry systems.

The aim of the farm-animal welfare movement, then, is not vegetarianism (the vegetarians have been around for decades and comprise only a minority that is no real threat to farmers), but rather to get the farm-animal industry to recognize that there are scientifically and ethically valid welfare issues that need to be addressed.

"Eating humanely" is one way that those of us who are committed to the humane movement and concerned about farm-animal welfare can make a significant personal contribution. I believe that it is morally acceptable to kill other living things—plants, animals, and their pests and parasites—in order to obtain the food that is essential to sustain our own lives, *provided such practices are ecologically sound and there are no viable alternatives to killing animals for food.*

The first principle in eating humanely is to ensure that the food that we do take from the plant and animal kingdom causes the least possible ecological destruction and waste of nonrenewable resources. In other words, we must try to reduce the rate of resource destruction and depletion. The use of regenerative (organic) agricultural practices is one good example of a practice that minimizes entropy. In this method, various crops are rotated so as to enrich rather than deplete the soil.

However, the large-scale agricultural practices in use today are accelerating entropy. Agricultural resources, including land, water, and plant crops, are being squandered in the highly inefficient overproduction of animal protein in the form of beef, pork, poultry, eggs, and dairy products. Until the production of such farm-animal produce is put on an ecologically sound basis, the first priority of eating humanely must entail greatly decreased consumption of all such foods. Farm animals can be part of the ecology of a more sound, regenerative agriculture, but today they are raised on large-scale factory farms. So a reduction in the production or consumption of farm-animal produce would mean a reduction in the number of animals being kept on these factory farms, and therefore a reduction in animal suffering and stressful overcrowding. Therefore, I recommend this dictum as a watchword: if you must eat meat, eat less of it.

The second principle of eating humanely entails selectively consuming only those farm products that come from animals that have been transported and slaughtered humanely. Currently, it is not possible to discover at the grocery counter whether the meat and poultry products on display came from animals that have been transported and slaughtered humanely. The Twenty-eight Hour Law gives protection only to animals transported by rail; today, most are transported by road, and humane standards for truck transit vary widely. There is a Humane Slaughter Act on the books, but poultry are excluded from protection, and inhumane treatment is not uncommon, especially in small, local packing plants that are not federally inspected.

So, in order to eat humanely, many personal decisions and sacrifices will probably have to be made. If you cannot be sure that the lamb, veal, beef, pork, bacon, ham, and chicken come from animals that have been transported, handled, and slaughtered humanely, you *can* make the conscientious decision to eat only products from animals that stay on the farm and are not transported and slaughtered: namely, eggs from laying hens (ideally infertile eggs from free-range hens) and milk, cheese, butter, and yoghurt from dairy cows.

A third principle of eating humanely is to endeavor to buy only those products from animals that have been raised under humane conditions and received good care during their growing period prior to slaughter, or during their productive lives on the farm. Again, one has no direct way of ascertaining in the grocery store which products satisfy these criteria. Until livestock and poultry producer associations establish and enforce their own humane codes* and humane labeling on all produce, to help the conscientious consumer select accordingly, you can, to some approximate degree of accuracy, rely on the following "humane scale" for selecting produce. This scale is of necessity somewhat arbitrary, since it is quite possible that some of the pork or eggs for sale come from relatively humane farm operations.

For assessing your eating habits from another perspective, the pyramid that follows the "humane scale" is based on many considerations, especially the relative prevalence of the various types of intensive factory-farming methods currently used for each species and the amount of time the animals spend under such conditions during their lifespan. The efficiency of conversion of feed into animal protein—i.e., high resource utilization—is the one mitigating factor in considering the humaneness of egg production, but this

*Along the lines of the American United Egg Producers' Association, and the British and European Economic Community Farm Animal Welfare codes.

Product	*Conditions*
LESS INHUMANE*	
DAIRY PRODUCTS (MILK, BUTTER, CHEESE, ETC.)†	Dairy cattle are the least intensively raised and confined of all farm animals, although there are some large-scale dairy "factories," especially on the West Coast.
TURKEY, DUCK AND CHICKEN	Animals have some freedom in deep-litter sheds and have a relatively short life, but conditions are often not conducive to animal well-being.
BEEF	Even though cattle are raised on grass/rangeland (where overgrazing and impact on wildlife habitat are problems), most are finished on grains and legumes in feedlots, a questionable use of natural resources and plant protein. Such diets are stressful. Furthermore, beef cattle are subjected to hot-iron branding and are castrated and dehorned without anesthetics.
LAMB AND MUTTON	While the majority of animals are not subjected to intensive confinement rearing, indiscriminate predator control in western states raises serious ethical and ecological concerns.
PORK, HAM, AND BACON	Many (but not all) are raised and finished in total confinement, and breeding sows are often subjected to unnecessary privation, confined in stalls, or tethered to the ground by a short chain.
EGGS	Most eggs come from battery-caged hens, the birds usually being extremely overcrowded to maximize profits.
VEAL	"Fancy" or milk-fed veal is usually from calves raised alone in narrow crates that severely restrict their freedom of movement.
MORE INHUMANE	

*No system of animal production (including transportation and slaughter) can be considered absolutely humane. In essence, some methods and systems are less inhumane than others, rather than being humane per se.

†While dairy products are graded as being less inhumanely derived than other farm-animal produce, this is not meant to imply that the dairy

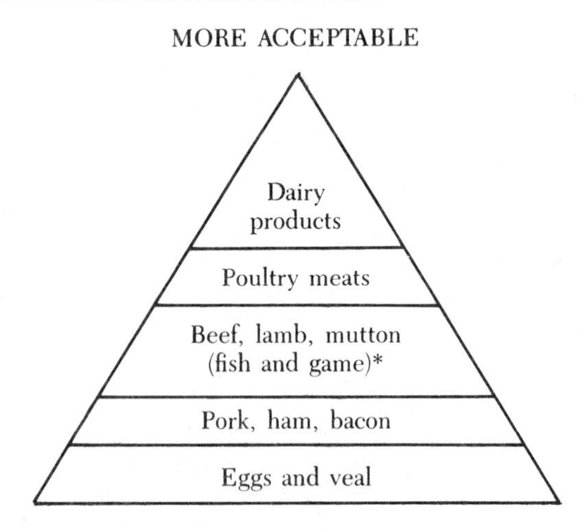

benefit is offset by the length of time that hens must spend confined in battery cages. Pork, ham, and bacon are scaled higher than veal, because not all pork producers have adopted intensive factory systems.

It is possible to be even more selective in choosing which farm-animal produce has been raised humanely by actually visiting local producers. In many towns there are often

---

industry has no serious welfare problems. Dairy cows, especially in large factory herds, are being pushed beyond their limits: high-energy diets cause stress, acid indigestion, and laminitis. These so-called "production-related" diseases are compounded by related metabolic disorders such as ketosis and fatty liver disease, and by management- and environment-related diseases such as mastitis and lameness, all of which shorten the life of the cow to around four to five years, by which time they are "spent." These problems are likely to intensify once the dairy industry starts to use bovine growth hormone to boost productivity even more. Since, each year, the industry overproduces, such treatment cannot be justified: it is a consequence of the economic treadmill that drives this and other agribusiness industries at the expense of animals' welfare and of smaller dairy farms that must either "get big or get out."

*See p. 127 below.

health-food stores that supply organic (drug-free) meat, poultry, dairy products, and eggs, in addition to local farmers' markets. You can get the names of the producers from these retailers and arrange to visit their farms. (Local humane societies could also help by drawing up a list of humane farms and thereby aid conscientious farmers to build up their business.) The following list provides a scale for selecting the more humane farming systems that you may well be able to find in your area:

BEEF — MORE HUMANE: free-range rearing and finishing (fattening) on open range or pastures with shade trees; open corrals with shade and shelter protection in winter; half-open barns with access to outside corral.

LESS HUMANE: feedlot corrals with no shade or shelter; overcrowded, closed-in sheds for finishing.

VEAL CALVES — MORE HUMANE: raising and finishing in group pens, ideally with straw bedding; or raising for the first 5–6 weeks in separate stalls and then finishing in group pens.

LESS HUMANE: raising and finishing in single stalls or pens in which they cannot walk or turn around.

DAIRY COWS — MORE HUMANE: open pastures with shade trees, or open barns with free access to stalls and an outdoor corral.

LESS HUMANE: outdoor corrals without shade or shelter; continuous restraint in separate stalls, usually in winter.

SHEEP — MORE HUMANE: free-range rearing and finishing on open range or pastures, with good shepherding to reduce need for predator control; finishing in half-open barns or sheds with access to outside open pens.

LESS HUMANE: feedlot corrals with no shade or shelter; finishing in overcrowded, closed-in sheds.

POULTRY* — MORE HUMANE: well-insulated and well-ventilated buildings with birds free on floor covered with dry deep-litter (often chopped straw or corn cobs), and with birds hav-

*Chickens are sometimes slaughtered at a few weeks of age and sold as "Cornish hens." Since these animals live for a shorter period than regular broilers and older birds sold as "roasters," it can be considered more humane to select these smaller birds that have been subjected to less stress during the course of their short lives.

ing easy access to food and water; adequate space for the birds to be able easily to avoid others.

LESS HUMANE: poorly insulated and ventilated buildings that become overheated in the summer and the air highly ammoniated and/or dusty, especially in winter; birds being allowed 1 square foot or less of floor space each, especially when they are close to slaughter age.

LAYING HENS

MORE HUMANE: free-range in well-managed pasture with access to a shed or house with perches and nest boxes, with straw or other suitable floor litter; or a covered or half-open barn or aviary with perches, nest boxes, and suitable floor litter in which birds are not overstocked.

LESS HUMANE: battery cages, of which there are many different designs. Some designs are more humane than others: ideally, each bird should be allowed at least 100 square inches of floor space.

SWINE

MORE HUMANE: well-managed open-pasture finishing, with shade and shelter; open-front and modified open-front buildings that are well ventilated, give good protection from climatic extremes, and are not overstocked. Straw bedding is a plus.

LESS HUMANE: total-confinement buildings that are poorly insulated and ventilated, in which the pigs are given 8 square feet or less of floor space each up to slaughter age; overcrowded battery cages for early-weaned piglets.

SOWS (when pregnant and between litters)

MORE HUMANE: outdoor pastures; well-insulated and well-ventilated individual houses or arks with shade and shelter; well-managed group pens in confinement buildings.

LESS HUMANE: total-confinement buildings in which sows are kept in narrow individual stalls or tethered to the ground.

It should be strongly emphasized that the humaneness of any husbandry system is also influenced by the attitude of the persons tending the animals and the quality of care and attention given to each animal. While an intrinsically inhumane, restricting, depriving, or overcrowded environment for the animals can only be aggravated by indifferent human attention and barely improved by careful attention, the quality of human care does play an extremely important role in those systems designated as "more humane." Basically, the more humane, less automated, and less industrialized systems depend greatly upon refined husbandry skill and high-quality human care. Such a system is qualitatively different from the management techniques needed to operate a highly automated factory system. But even on factory farms, attentive, concerned managers can do much to enhance animal well-being.

Fish is a popular consumer food and, while fish are *not* humanely killed, ocean fish are at least able to live a natural life. But problems of overfishing, resulting in severe depletion of stocks, and contamination of fish produce with harmful pollutants (especially heavy metals in the larger fish species and in invertebrates such as clams and oysters) make the consumption of fish seem a less sensible alternative than ranch-raised fish. But the latter are often overstocked and given medicated feeds to offset the stresses of confinement rearing. Furthermore, since Japan and Iceland are major exporters of fish to the American market, and these countries insist on continuing to harpoon whales, fish products from these sources (which usually bear a label of their country of origin) are unacceptable.

Aquaculture, or the raising of seafood in closed, controlled systems, would seem to provide hope for an ecologically sound, humane source of protein. Unfortunately, however, the same factory methods we have been encountering on the farm are appearing here.

"Aquaculture 1983" was the title of a five-day symposium and industry exhibit held in Washington, D.C., on January 9–13, 1983, sponsored by World Mariculture Society, Catfish Farmers of America, Fish Culture Section of the American Fisheries Society, U.S. Trout Farmers Association, Shellfish Institute of North America, and National Shellfisheries Association. While ecologists, economists, futurologists, and others have touted the virtues and potentials of intensive fish and shellfish farming, this growing industry may become blighted by the same problems that have come to afflict factory farming of crops, livestock, and poultry.

Industry exhibits told the story—there were displays on herbicides and algicides to control the proliferation of plant life in overstocked and polluted fish ponds, and aeration systems to help alleviate pollution from fish excrement and rotting food in the water. Antibiotics such as tetracycline and sulfonamides were promoted for incorporation into feed, along with other drugs to control fish parasites and fungal infections. And a variety of autogenous bacterins (vaccines) were also marketed to help combat disease. One industry exhibitor even admitted that all this was necessary because, just as in agriculture, the use of monocultures (raising of a single species) is ecologically unsound and creates disease problems. Another spokesman added that all these exogenous agents are necessary because the fish are crowded and so are under stress and therefore more prone to disease. Bacterial resistance to some antibiotics has already emerged as a recognized problem.

In sum, aquaculture is now on the agribusiness treadmill of increasing dependence on technology and drugs (thereby providing a lucrative business for support industries, especially the chemical and pharmaceutical industries), in order to rectify intrinsically unsound husbandry practices. The possibility of organic and humane aquaculture, without overstocking and overuse of drugs, fades into improbability as the values and economic structure of the rest of agribusiness begin to saturate this fledgling industry.

In assessing the importance of aquaculture to add a source of protein to our diets, we must weigh the fact that we are now feeding an estimated fifty percent of our fish catch to livestock.[11]

Wild-game meat, such as venison, ranch-raised pheasant, and antelope, may be placed in a comparable position on the humane scale as beef, lamb, and mutton. Humane slaughter cannot be assured, but at least wild game, until the time of death, have lived a relatively natural existence compared to confinement-raised livestock. But again, predator control to reduce hunter competition and other wildlife mismanagement practices to increase populations of desired game species are critical negative factors that must be considered. While ranch-raised game may suffer from overstocking and poor management, some species, such as deer and antelope, convert feed more efficiently than do cattle. Thus, in some areas, wild-game ranching may be ecologically more sound than raising cattle or sheep.

Closely related to the concern of humanely feeding ourselves is the question of how we feed our pets. Many humanitarians and vegetarians own meat-eating cats and dogs. The pet-food industry, a subsidiary of agribusiness, is a $4.25 billion annual enterprise which utilizes a variety of agricultural products and food-industry byproducts, including cereals, soybeans, milk, whey, eggs, chicken parts, meat byproducts (liver, kidney, tripe, animal fat, bone meal) dairy byproducts (whey and cheese meal), and fish meal and various fish parts. Pet food also contains various coloring

*Some of these pet-food dyes are responsible for turning white- and silver-colored cats and dogs into orange and brown animals. The source

agents,* chemical preservatives, flavoring, and vitamin and mineral addditives. Many of the well-known brands of commercial pet foods provide scientifically formulated and balanced diets, but some veterinarians are beginning to question just how healthful such processed foods actually are for pets.

The myth that a high-meat diet is essential for human health has proliferated to pet-food advertising. While admittedly cats, and to a lesser extent dogs, are, as carnivores, more meat-dependent than we are, a high-meat diet can still jeopardize pet health. Some pet-food companies have even advertised the benefits of an "all-meat" diet for dogs and cats: this promotion is a travesty of nutritional science. One company, Alpo, has even pushed the notion that emotionally distressed dogs need a high-protein diet. The company attempts to shore up its assertion with scientific data showing that a high-protein diet helps sled dogs cope with the physical stress of competitive racing. However, there is little connection between emotional distress and physical stress; while a high-meat diet may help mitigate the effects of the latter, it will not alleviate the distress of the former, such as the dilemma of a dog that is left at home alone all day while its owners are away at work. In fact, an all-meat diet can cripple growing puppies and kittens because there is too little calcium in this kind of food, and a high-meat diet can kill older dogs by uremic poisoning. Likewise, if they are not fortified with vitamins and other essential nutrients, all-fish-based diets (which are widely marketed for cats) can cause serious nutrition-related health problems. Given the readily available abundance of animal protein from our overproductive factory farms, from the astronomical losses that accrue during live-

---

of a pet's drinking water introduces another factor: the color is derived from the copper that is leached from water pipes, as the tapwater in many regions increases in acidity as a consequence of acid rain.

stock transportation, and from the diseased and injured parts of animals condemned for human consumption, it is only natural that pet-food companies promote the idea that pets need a high- or all-meat diet.

But the terms "all-meat" and "meat byproducts" are misleading. Labeling which indicates the percentage of protein in the food is of limited value: many designated meat byproducts, while they are indeed animal proteins, are not actually meat. These include ground-up lungs, udders, intestines, skin, and even hoof and horn, which are highly inferior proteins since they are not easily digestible. On the positive side, however, these animal products do provide some roughage, even though the owner is being misled by the labeling into believing that the food contains only *meat* byproducts.

The Pet Food Institute, which represents the pet-food industry, petitioned the Food and Drug Administration to approve proposed changes in the regulations that govern what information manufacturers must include on their labels indicating the constituent ingredients of each product.[12] Under the suggested new regulations, corn husks and peanut shells would be listed simply as "vegetable fiber," hydrolyzed poultry feathers as "processed poultry protein products," cheese rinds as "cheese," and ground bones as "processed animal protein." Furthermore, the actual ingredients in each can of product would not be ascertainable, since class (or category) names of some ingredients would be allowed—"cereal grains," for example, could mean rice, barley, or wheat. Such vagueness in labeling could cause serious problems for those pets who are allergic to certain foods (wheat, which can cause epilepsy in dogs), since the pet owner would not know whether the harmful ingredient was present or not. Furthermore, labeling animal byproducts that are actually of little or no nutritional value as "protein" constitutes a practice that is not only misleading to the public; it can also be detrimental to animal health. To stop the proposed changes in labeling regulations, the Humane Society of the United States and the American Veterinary Holistic Medicine Association voiced their opposition.

This recent move by the pet-food industry (which they claim will save themselves and consumers $200 million per year) may be motivated by three factors. First, the cost of quality ingredients continues to increase steadily. Second, the industry trend of "lowest-cost feed formulation" leads to a downward spiral of deteriorating products, in which competitors strive to undercut each other by manufacturing palatable but ever lower-quality food with cheap ingredients. (This is a serious problem with generic pet foods.) Third, these cheap and readily available ingredients are byproducts of the highly diversified agribusiness/food industry; the industry therefore hopes to profit by dumping its waste products into pet-food subsidiaries, rather than using them as organic fertilizer.

Veterinarians are finding that many pet-health problems, especially those affecting the skin, are partially alleviated or totally cured simply by taking the animal patient off all regular processed commercial pet foods. Recently, generic dog foods have been linked to a variety of health problems, notably poor growth and disease susceptibility in puppies and outbreaks of severe skin disease in dogs caused by zinc deficiency. The latter problem may be associated with the high cereal content of pet foods (since phytases in cereals can block the absorption of such essential trace elements). As the cost of ingredients increases, more cereal and cereal byproducts are put into pet foods. This is already a problem for larger breeds of dogs (who should be fed several small meals each day rather than one large one in order to prevent bloat).

Many commercial dog foods contain too much calcium and phosphorus, which can cause a variety of problems, including zinc-deficiency dermatitis, bloat, and hypothyroidism. This excess calcium comes from calcium-salt supplements and bone meal. Cats, dogs, and humans suffer the consequences of too much salt in prepared foods, with hypertension and increased risk of subsequent cardiac and renal disease. Excess protein is linked with renal disease (and possibly arthritis), the harm being increased by excess sodium and phosphorus. Excess phosphorus is linked with hyperparathyroidism. And cats suffer from cystitis and urinary blockage from the excess of magnesium in commercial pet foods. Commercial foods too high in fats and/or carbohydrates are linked with skin problems and obesity. Bone meal contains lead, an immunosuppressant, and the various parts of animals—especially the fat—used in pet foods, contain toxic pesticide and other agrichemical residues* and their unknown, unidentified metabolites which may be even more harmful; hormones and other drugs, including arsenic and antibiotics that are given to farm animals; and industrial pollutants such as PCB, PBB, and cadmium.

Heat processing of commercial pet foods destroys heat-labile enzymes and lysine, and essential amino acid.

Food supplements, such as zinc and vitamin E, may help compensate for deficiencies in pet foods but can cause nutrient imbalances in bioavailability and bioeffectiveness. Thus, fortification and enrichment of deficient commercial pet foods with trace minerals and vitamins is only a limited improvement. Agrichemical and industrial pollutants, processing contaminants, processing effects on nutrients, pro-

*Many of these are suspect immunosuppressants, teratogens, mutagens, and carcinogens. The 150 and more genetic diseases in purebred dogs may be linked with these and with inbreeding which may increase susceptibility to such environmental food-contaminants.

portions of basic nutrients (protein, fat, and carbohydrates) and chemicals added by processors (coloring, flavoring, antioxidant preservatives, stabilizers, extenders, etc.) pose other problems, few of which have been studied; those that have, indicate that physical and psychological (behavioral) problems can arise from these other factors, including cancer, immunosuppression, nervousness, hyperactivity, allergies, and diabetes. In sum, we see in pets who eat food that is similarly contaminated and adulterated precisely those diseases we see in humans.

Since dogs and cats are basically carnivores, it is not advisable to attempt to impose a strictly vegetarian diet on one's pet, as some concerned pet owners have done without some expert advice. The smaller breeds of dogs, puppies, older dogs, and pregnant and lactating bitches may not thrive unless their diets include some high-quality protein, such as dairy products. (Some dairy products, which are high in lactose, can cause digestive problems in cats and dogs, so an alternative source of animal protein may be advisable.) Cats can be said to be more carnivorous than dogs (which are relatively more omnivorous), in that they generally require a full complement of amino acids daily. This necessitates providing a cat more animal protein than you would for a dog of comparable size.

The consequences of how we feed our pets raise the same specter of health, ecological, and humane concerns as those that are germane to our own eating habits. On the other hand, we can apply the same basic principles of eating humanely in the feeding of our pets that we use in determining our own diet.

Some critics of Western civilization have noted that the average pet in America eats better than the average child in Third World countries. This should not make cat and dog owners feel guilty, but at least they need to recognize that keeping carnivores as pets is a costly luxury and that a reduction in the overall population of these animals would be prudent, since there is a serious overpopulation problem (some eight to nine million unwanted cats and dogs are destroyed each year in American animal shelters).

6

Measuring the Results

THE question that remains to be asked about current American farming is, "Does it work?" That is, are we getting the most plentiful, varied, and nutritious food in the world at a price we can afford? And, if so, has this come about because of or in spite of the methods we use to produce it?

We are told that a single farmer can feed 78 people and that Americans spend only 17 percent of their disposable income on food, the lowest proportion in the world. Behind these figures is another reality. While there are only 2.5 million farmers (and this figure is declining every day), another 22 million people are employed in the off-farm food-production and -delivery system, America's biggest industry. As for spending only 17 percent of our income on food, this figure is way off for low-income families.

In 1967, the USDA report *Economics of Scale in Farming* revealed that the small one-to-two–person farm is just as efficient as—and often more so than—massive corporate farms. Studies since then have confirmed that assessment. The most efficient production unit (in terms of *output per acre*) has repeatedly been shown to be quite small, although larger farms can produce more food *per unit of capital and labor.*[1] Some examples: pigs fatten more slowly and hens lay fewer eggs when they are overcrowded, but more meat or eggs are produced per unit of space and volume of food consumed. North Carolina State University extension economists, in a 1979 survey, found that the average returns to management are higher on medium-sized (65–69 cow) herds, lower production per cow being the most common reason for lower returns in larger herds of 100 or more.[2]

L. Tweeten finds:

> Although the optimal size of a farm, if there is one, varies widely and no one size fits all conditions, the size of farm consistent with increased well-being of society as best measured with our crude tools is neither a small nor a very large farm but rather is a moderate-size family operation. In short, the optimal size farm to increase well-being as best that can be measured appears to be the typical size commercial farm of today, approximately $100,000 in sales and $1 million in assets. The nation could currently support approximately 1.2 million such farms, or twice the existing number of family-size farms.[3]

Professor Stanley Curtis of the University of Illinois has been a strong supporter of confinement systems. Yet in his own textbook,[4] he cites several studies that show no economic advantage to intensive confinement systems. He reviews studies comparing the costs of different housing systems for dairy cattle, cows, beef cattle, swine, and poultry. In all instances, the costs of production are essentially the same. The reason for this is that the costs of labor saved in confinement operations are offset by increased capital costs for construction and maintenance, plus the need for costly drugs and other health measures to maintain productivity. The one exception, however, to this general finding was in egg production. It was found that there would be on average a ten percent increase in the cost of production of eggs from hens housed in open barns and sheds instead of in battery cages. However, this increase would result in an increase in the retail price of eggs of somewhat less than five cents per dozen.

In a detailed review entitled "Is Factory Farming Really Cheaper?"[5] Jim Mason shows very convincingly that while the real prices of animal products have declined since the end of World War II, it is not true that these lowered prices are a consequence of the adoption of intensive-farming practices. "Other factors, such as chronic overproduction in the milk and egg industries, grain surpluses and, especially in the broiler industry, mechanization of processing and fierce price competition among brand leaders, have been the main forces in keeping prices down."

In Great Britain, Quantock Veal has abandoned the raising of milk-fed veal in single pens or crates, opting instead for group pens. This has halved the cost of raising each calf. Economic comparisons of open-front, modified open-front, and total-confinement swine-production systems in the United States have shown clearly that there are no clear-cut economic advantages to producers in any one system over any other.[6] Professor Curtis concurs: "Rather large differences in housing type do not necessarily affect food-animal health and performance. Indeed, acceptable performance is possible in a wide range of accommodations—provided the facility and animals are managed adequately."[7]

The standard measure of efficiency is the ratio of the returns we harvest from a system compared to what we put into it. Chapter 2 has detailed the direct expenses of farmers for such purchases as drugs and veterinary care, fertilizers and pesticides. For the period studied, we have seen that drug and veterinary costs outstripped the general rise in farm production costs.

Prohibiting drugs in animal feed would save about $2 billion a year, but losses due to disease or the slower growth of the animals might outweigh the savings until animal-husbandry practices and nutrition are improved. The projected impact of a proposed ban on adding antibiotics to livestock feed has been summarized in the USDA's *Agricultural Situation*.[8] It is calculated that broiler output would be cut by eight percent, turkeys by six percent, and hogs by five percent within a year after implementation of the restrictions. Beef production might increase slightly, due to lower feed costs and a predicted increase in steer prices. The production of eggs, milk, and lamb would not be significantly affected. It is calculated that it would take five years for the production increases that would eventually result from higher prices to eliminate most of the immediate, negative economic effects of this ban. The report notes that

> restrictions on drug use could cause some profound changes in the structure of the livestock industry. For example, if low-level dosages of antibiotics are essential for successful confinement rearing of a large number of animals, banning their use would have dramatic implications for poultry and hog production.

The above figures of predicted decreases of eight percent for broilers and five percent for hogs that would result if antibiotics were to be withdrawn serve to underline the critical role in buffering the unhealthy effects of stressful rearing conditions that is played by the use of protective drugs. A ban on drugs in the animals' feed could lead to a revolution in husbandry practices.

The American Council on Science and Health is a new "expert" panel of scientists dedicated to furthering the interests of agribusiness. Virgil Hays, an animal nutritionist and ACSH spokesman, claims that banning penicillin and tetracycline in feed would cost drug companies, livestock producers, and consumers plenty. His estimate: $3.5 billion.[9] But note that this figure includes the loss of sales by drug companies, which is a *bonus* to the producer and would have a positive effect on consumer prices.

In England, the use of nitrogenous fertilizers has increased eightfold since 1900, but the per-acre yield is only 1.5 times what it used to be—an increase that may not reflect the "benefits" of synthetic fertilizers but, rather, gains from other improvements in farming.

An annual 1.4 billion pounds of pesticides are applied in the United States, at a cost of $4.2 billion, a *one-thousand percent* increase since 1947. But, as Professor David Pimental has shown, crop loss caused by insects has almost doubled (from 7.1 percent of total crop value in the 1940s to 13 percent in 1980).[10] Pimental estimates that the cost of applying pesticides to U.S. cropland is about $2.5 billion a year, and the crop loss is about $10 billion.

> Food surpluses for domestic and export use could be larger if insects, pathogens and weeds did not destroy an estimated 33% of our potential crop production. This loss occurs in spite of the use of pesticides and other pest control strategies. Obviously, a zero loss of crops to pest attack would be impossible to achieve because of economic, environmental, biological, and public health constraints. Reducing the 33% loss by about half, however, may be a reasonable goal.
>
> Taking into account that 17% of U.S. crops acreage is treated with herbicides, 6% with insecticides and 1% with fungicides, crop losses without pesticide use and employing some alternate controls were estimated to increase from 33% to 41%. Crop losses, based on dollar values, thus increased by about 8%.
>
> The increased human food energy loss without pesticides was estimated at 4%. Based on this estimate, no serious food

> shortage would occur in the United States if pesticides were withdrawn, because most staple foods, such as wheat, would not be greatly affected by the withdrawal of the chemicals. However, the production of certain fruits and vegetables, such as apples, peaches, onions, and tomatoes would be greatly affected. . . .
>
> The environmental impact of pesticide use, including human pesticide poisonings, is estimated to cost the nation at least $3 billion annually. Taking these external costs into account in a benefit/cost analysis of pesticide use in agriculture, the dollar return is $3 per $1 invested in pesticides.
>
> This analysis of the benefits of costs of pesticides is but a preliminary assessment of a vital issue, one that deserves greater study so that safer and more effective pest control strategies can be developed. Of particular importance in subsequent analyses is the need for a more extensive, indepth assessment of the external costs of pesticide use. Data is presently scanty and it is difficult to put a dollar cost on some of the effects of pesticides. Such information is necessary if we are to make sound decisions concerning the benefits and costs of pesticide use in U.S. food production.[11]

Most of the inputs into factory farming are made of or use fossil fuels. The American fossil-fuel–based agricultural system is extremely inefficient. To feed the world, it would use sixty percent of the world's energy supply and would exhaust the earth's known petroleum sources within thirteen years.[12] Further, seven times more fossil fuel energy is required, on average, to produce one pound of meat than one pound of grain protein, and sixteen times more labor.[13] Our dependence on fossil fuels is such that a ten-percent reduction in oil supply would result in a fifty-five-percent increase in the price of vegetables and fruit.[14]

The farming of animals uses much more energy than the farming of grains. Dr. W. L. Roller and his colleagues have shown that only the best-run animal factories achieve a 34.5 percent return as food on their total fossil-fuel energy investment. In contrast, even a poor crop of soybeans or corn would provide a return of 328 percent.[15]

Professor Pimental recommends that we reduce some of our energy consumption by raising chickens instead of beef and grass-feeding all livestock (dairy, beef, and sheep). This, he points out, would save "up to 30 gallons of oil per person . . . each year, or one percent of the entire energy consumption of the nation." Only one-half the energy required to produce beef protein is needed to produce broiler protein. In a grass-feeding system, sixty percent of the energy we now use to grow grains could be saved. (He further notes that more grain, currently valued at $20 billion, would then become available for export.)[16]

On the other hand, Pimental concludes that if forage were to entirely replace the grain now fed to U.S. livestock, animal-protein production would decrease by nearly one-half. And although the animal protein that would then become available per person would be nutritionally adequate if all had equal access to it, the poor would be least able to compete in obtaining it.[17]

The consequences of foraging to natural habitats and wildlife, however, are devastating, as we have seen (pp. 53–56). Extensive grass-feeding replaces one ecologically unsound practice with another.

Tractors and other heavy farm equipment constitute still another drain on our fuel resources. Horses and mules as draft animals on American farms were almost completely replaced between 1920 and 1950, yet this was accompanied by "inconsequential" increases in corn and wheat yields.[18]

All animals are not equally efficient converters of the energy they consume.

Under current production conditions in the United States,

> the efficiency with which protein is produced in grams per Mcal of DE (digestible energy: dietary gross energy minus fecal gross energy) ingested is as follows in various food products: milk, 12.8; broiler, 11.9; eggs, 10.1; pork, 6.1; and beef, 2.3. These outputs increase as the intensitivity of production (mainly energy input per unit of time) increases because the energy cost of maintenance becomes a gradually smaller proportion of the total energy input.
>
> If, with the increasing human population, the supply of grain, oilseed, and pulse feedstuffs for livestock were to become limited, it would be necessary to determine which animals are husbanded for food production. In terms of protein output per unit of concentrates ingested, dairy cows producing milk are the most efficient users of feed energy. Among farm animals, swine have the most precarious position as food producers; pork is relatively low in protein (usually about 13.5% of conventional slaughter weights), pigs are competitive with man for food or for land on which to grow man's primary foods, and the swine production enterprise requires a high fossil-energy subsidy. Despite the low output of protein per unit of DE ingested by beef cattle, beef cattle produced under an all-forage system require very little fossil-energy subsidization and, because of the abundance of cellulose, the future role of beef cattle and other ruminants as food producers is assured.
>
> Because of the continuing depletion of the world's fossil fuel, much research is needed to develop low-energy consuming systems of animal production. Under conditions of intensive management in the United States, the fossil energy demand of protein production is least in milk, intermediate in pork, and greatest in beef. Under an extensive system based on all-forage diets, the energy subsidy is reduced by as much as one-half for both milk and beef production, even in the northern states.[19]

The huge loss of animals on the farm from disease and at the slaughterhouse from condemnation is another major element in America's food bill.

The measurable costs we have been looking at are those that are passed along to the customer in the retail price he pays. They are only a small part of what society is actually paying for our "cheap food." Chapter 2 has shown some of what we are spending indirectly in taxes for taking land out of production, price supports, tax credits, surplus storage, and a displaced work force.

Not showing up in these statistics is the havoc we are wreaking on our topsoil and water, and on the earth's rain forests and grazing lands (see Chapter 3). Pimental has pointed out that a fifty-percent reduction in consumption of meat and other animal products could save half the energy, mineral resources, and land, and one-third the water, used in animal production.[20]

Nor does the toll in animal suffering appear here, yet measures to reduce suffering are rejected for economic reasons. Some immediate humane reforms can be instigated *without* increasing costs by much except, initially, those required to ameliorate veal-calf production. In fact, most of the necessary changes will be likely to increase profits (except to pet-food companies, who benefit from cheap, damaged carcasses). Prime candidates for reform include the method of disposal of unwanted male chicks in hatcheries; poor conditions of transportation of livestock and poultry; methods of forced molting of laying hens; dehorning and castrating of cattle; debeaking of chicks; and crate-raising of veal calves (and veal is a wasteful dietary luxury compared with other protein sources such as eggs, chicken, and dairy products).

Economics are not, however, the sole measure that should be considered. A report, "Animal Welfare in Poultry, Pig, and Veal Calf Production," delivered by the Agricultural Committee to the British House of Commons in July 1981, concludes:

> We do not accept the contention, frequently stated or implied, that the public demand for cheap food decrees that the cheapest possible methods of production must be adopted. . . . Society has the duty to see that undue suffering is not caused to animals, and we cannot accept that that duty should be set aside in order that food may be produced more cheaply. Where unacceptable suffering can be eliminated only at extra cost, that cost should be borne or the product foregone. On the other hand, all methods of domestic livestock rearing entail some loss of freedom, and where an imperfect but not unacceptable system can be improved only at disproportionate cost, it may be unreasonable to insist that this be done. Once again a balance has to be struck, and this can only be done in the light of subjective judgment; but our emphatic view is that the welfare of the animals must come first.

The human factor is addressed in a study of European farming practices:

> It is also apparent that nearly all the improvement in efficiency associated with size of operation can be substantially achieved by businesses which are very small compared, in terms of labour force, with firms in most other industries. Little evidence is found that any appreciable increases in efficiency are to be gained by enlarging beyond the point where the farmer remains in very close contact with his employees. The impersonality of big business which has been blamed for depriving employees of their sense of identity or self-respect and alienating them from their employers has no significant place in British farming.[21]

Jim Mason sums up the cost-effectiveness debate:

> The financial benefits of factory farms are exaggerated, and . . . they produce unhealthy animals and poor-quality products: to offset these effects factory farmers must employ an arsenal of antibiotics, hormones, drugs, chemical additives, coloring agents and other substances that may threaten human health. When one considers the potential magnitude of these

health problems and the social costs of dealing with them, the food produced by factory methods may well be too expensive—regardless of its price at the market.[22]

The "progressive efficiency" myth of agribusiness and industrialized farm-animal production is exploded by T. Bender:

> Efficiency is a myth, and a myth that we cannot afford. It is a false measure of the efficacy of a process, or its power to produce intended effects. Ecologically, efficiency is equated with instability, low number of species and relationships, and ease of disruption. The loss in efficiency of a particular process necessary to make it an integral and non-disruptive part of an ecosystem is the energy necessary to sustain the complex relationships and alternative pathways which give that process the flexibility to adjust, react, and sustain itself under varying conditions.
>
> Efficiency is an artificial effectiveness that can be sustained only in intensely uniform and artificially maintained conditions. It is an unusably narrow measure that measures only directly attributable costs, and only those costs expressible in terms of dollars. It has no way to compare with processes based on other than purely economic criteria, and cannot measure either good or bad results inexpressible in dollars or not directly attributable.[23]

Whether we measure the cost of our food by the green stamps we get or include the vastly greater external expenses we have been enumerating, are we getting what we pay for? The food we consume is adulterated with hundreds of chemicals and drugs whose effects are not yet known. We are spending huge amounts to research and cure diseases that may, at least in part, be caused by what we eat, and we are making laboratory animals suffer in the process.

Chapter 3 has set forth some of the myriad health problems arising from chemical and bacterial contamination of our food. But improving these conditions is measured in terms of the cost to the food industry—not to society at large. For instance, the costs of salmonella poisoning to society, in terms of lost wages, hospital costs, and death, are greater than the savings the meat industry would enjoy by eliminating it from poultry (as tuberculosis has been eliminated from cows) and by reducing fecal contamination of meat products, even though this results in food poisoning and the spread of bacterial resistance to antibiotics. So nothing is done.

Dow Chemical Company maintains that "everything natural or man-made is comprised of chemicals and man has adapted through time to live exceptionally well with both natural and man-made toxic elements."[24] This rationalization and denial of the hazards of manmade chemicals ignores the fact that people are exposed to thousands of hazardous substances and that adaptation is a gradual, evolutionary process. Measuring the effect of one chemical in laboratory animals does not measure the effect on the environment or its potentially harmful synergism with other chemicals in our own bodies. While over generations we and other organisms acquire immunity to naturally occurring poisons in plants, the scenario is quite different when, in one generation, we are faced with seventy thousand or more manmade chemicals, many of which have been shown to be harmful to living organisms.

Regulation and inspection are not doing the job.

Even without the contaminants and food additives, much of what we eat is unhealthy, containing often excessive amounts of the "four white poisons": sugar, salt, refined flour, and saturated fats.

Thus, while it is falsely claimed that American agriculture is the most productive in the world, in actuality it is probably the most polluting and wasteful. For instance, an estimated $32 billion is lost annually when produce, often only slightly damaged, is discarded by retail stores. Mechanical harvesting leaves a considerable percentage of some crops to rot in the fields, which leads to high populations of birds such as starlings and blackbirds, and these species then become "pests." The vast amount of fecal material from animal factories also poses serious waste-disposal problems: the runoff (along with fertilizers and pesticides from fields) causes water pollution and eutrophication. Pesticides poison wildlife and, through the runoff, contaminate the fish we eat and the water we drink.

The only way to break out of this vicious cycle is to make a dramatic cut in the current level of production and consumption of farm-animal produce, but this move would require a sacrifice that farmers should not bear. Rather, the consumers of their produce should shoulder the requisite burden, by paying such higher prices as are necessary to facilitate the transition to a more humane and ecologically sound agriculture. Most Americans can spend a bit more since, at present, they pay significantly less for their food* than the people in any other country in the world. This change will also help reduce the pressures on our dwindling environmental resources and help restore American agriculture to a regenerative, ecologically sound, and self-sustaining balance between production and consumption.

The religion of American agricultural efficiency has been sold to the Third World.

We encourage the use of tractors without any awareness of the overall ecology of cattle in other cultures. In many parts of the world, cattle are indispensable to human existence for draft work and as a source of protein in the form

*In *retail* terms: reducing the enormous indirect costs would offset much of the price increase.

of milk, butter, and cheese. Cow dung is also an essential fuel (though innovations such as methane-gas production from manure and use of the liquid remains as a pasture fertilizer may be ecologically more sound).

We emphasize higher crop yields through increased reliance upon fertilizers that are produced from petroleum. Human populations are encouraged to expand beyond the level that can be sustained by sole reliance on natural sources of nitrogen.

The introduction of nonnative species of crops forces Third World farmers to rely on the same chemical methods that are proving so disastrous to American agriculture. The "Green Revolution," which ostensibly aimed at alleviating world hunger by improving agricultural productivity in poor countries, has caused more harm than good. Agribusiness, aided by the USDA, has helped to spread ecologically unsound and inappropriate factory-farming methods to Third World countries and, along with it, a demand for meat and a dependence upon American technology, seed stock, fertilizers, agrichemicals, and animal-feed grains. Pharmaceutical and petrochemical companies profit greatly from sales of pesticides, antibiotics, and other agrichemicals, some of which—banned in this country because of their high toxicity—are nevertheless widely exported. Further, the benefits of increased production simply do not reach the needy, according to many analysts.

Likewise, we promote the introduction of domestic plants and animals at the expense of native species, without considering the alternative of farming *with* rather than *against* wildlife.

Without costly energy inputs, introduced domestic livestock would not survive. David Hopcraft observes, "Each year, major government and financial institutions pour millions of dollars into research and development projects, to the same thoughtless and ultimately disastrous end."[25] Dr. Hopcraft's research supports the notion that, where appropriate, wildlife ranching will be more productive economically as well as providing the sole solution to the problem of preserving natural habitat and indigenous wildlife. He has shown that 14.6 pounds of lean meat per acre can be produced by ranching gazelles, while only 3.9 pounds per acre are derived from well-managed cattle (there is only a 1-pound yield per acre with traditional stock-raising methods). Biomass of stock in pounds per square mile can be more than tripled if adapted indigenous species are ranched. More hides, of higher value, are also produced. Cost-effectiveness is greater for a wildlife-ranch system; with fewer overhead costs (for providing water, disease control, etc.), gross income can be three times greater than for a cattle-ranch operation. The meat of wild game may also be healthier for the consumer, since it is 90–99 percent lean and contains no saturated fats.[26]

It is ironic that, at precisely this time, the Third World is being encouraged to adopt Western systems of agricultural livestock production and to acquire a taste for meat. The economic, social, and ecological consequences of such a promotion may well be disastrous.

It is now well documented that the multinational corporations who, together with investment banks, comprise the American agribusiness system, have been major contributors to famine and the destruction of farmlands in Third World countries. By keeping world prices low, the United States puts in jeopardy Third World farmers who cannot compete with the cheap agricultural produce we dump on their local markets. These farmers are then forced out of business, and the countries involved become increasingly dependent upon American imports for their basic food needs.

Another way in which U.S. agribusiness with its global

reach harms other countries entails the establishment of corporate farms and plantations, often funded by the World Bank under the guise of economic development. As a consequence, peasant and native peoples lose their land and are forced to seek employment in the cities or on the corporate farms. These farms produce exportable cash crops (such as cotton, coffee, animal feedstuffs, sugar, and alcohol)—not food. Only the elite in these countries profit, while others starve. Serious degradation of agricultural lands occurs because sustainable and regenerative agricultural practices are replaced by monocultures which rapidly deplete the quality of the soil. When we ask if American farming works, we must ask if it works for the world—for that is its true scope.

It is clear that we need a new yardstick by which to measure progress and efficiency, its cardinal parameters and denominators being based upon sound scientific evidence and ethical principles derived not from analysis of narrow, short-term costs and benefits, but from a more holistic, egalitarian, equitable, and ecological perspective.

7

Toward a Saner Future

I MUST stress that I do not think the present state of affairs is the fault of the independent farmer, the animal-production scientist, or the veterinarian. The villain is ideological: agribusiness industries that try to tell us that "bigness is best," that we are getting cheap and wholesome food, that we and the environment can tolerate the damage being inflicted upon us.

It must be apparent that we cannot go on as we have been. We are killing the earth, killing the animals, killing ourselves—this is the true meaning of agricide.

It is not too late to change our ways. We must wrest control of American agriculture from those large corporations that produce and encourage the use of chemicals, the maltreatment of animals, and the monoculture of crops. They cater to their stockholders, not to society.

We must, as has been shown, reduce our intake of farm-animal products, for this is the root cause of much that is wrong. And we must change the way we raise those animals we do consume. Animals and the poor are now in competition for grains and legumes, and this need not be.

The guiding principle is that all meat products (except for the highly efficient broiler chicken) should be marketed only as byproducts of three primary industries: wool, egg, and dairy production. "Spent" laying hens and dairy cows can be utilized (as they are today) in canned soups, stews, and pet foods. Similarly, mutton can be derived from wool-producing sheep (and goat meat from milk-goats) that have passed their prime. In this scheme, the male offspring of these animals comprise a second category of animal-protein sources and, according to the availability of feed resources, could be killed at varying ages for human consumption as byproducts of the three primary industries. Also, since these industries do not entail the wholesale killing of animals and since only the animals' products (wool, eggs, milk) are used, problems associated with inhumane transportation, handling, and slaughter would diminish to relative insignificance.

Contrary to agribusiness's claims, a revolution in beef, pork, and lamb production would not mean that farmers would be forced out of business. Those whose primary product at present is meat could quickly diversify their farming practices within a few years. Such diversification would be in their own best interest, since even the most conservative long-term predictions indicate that specialized meat production is not cost-effective. And, given the unpredictability of grain harvests and world demand for grain and other crops, American agriculture is cutting its own throat in trying to maintain its export production and, at the same time, encourage domestic overproduction and overconsumption of meat. Since many developing countries are now rapidly becoming more self-sufficient through enhanced agricultural technologies, it is unwise for American farmers and the U.S. government to continue to place so much emphasis upon raising crops for export.[1] Promoting meat-eating, because of the inefficiency of grain utilization that is involved, also means inflated food prices for consumers. A dramatic reduction in the production and consumption of all "red meat" and an overall reduction in the production of eggs and dairy produce, with increasing emphasis on diversified farming to supply local markets with fresh fruits and vegetables (which will also greatly reduce transportation costs and spoilage), would mean that most farmers will be able to stay in business. Further, they will prosper once again and will be fairly rewarded for their good labors. Today, with few exceptions, this is not the case.

We must divorce agribusiness from the USDA and our land-grant universities and extension services, so that research and consumer education will follow the path of reason. Grants to spur research into alternative regenerative methods of farming are urgently needed—and a new mission for land-grant universities to develop ecologically sound and humane practices, rather than the drug-dependent, high-tech, capital-intensive systems that have been to the detriment of agriculture and society alike.

We must break the hold that agribusiness has on our national policies and fight for more logical tax codes and regulatory practices. The livestock industry's fears of restrictive legislation and bureaucratic oversight focusing on animal welfare are not without foundation. Enactment of premature legislation could be disastrous for both livestock and producers. More research is needed in some husbandry areas to define what is optimal from a welfare point of view and within an economically acceptable framework. At this stage, separate task forces need to be set up, composed of representatives of various groups within the livestock industry, along with consumer and humane representatives. These task forces should look into the hog, broiler, egg, milk, and veal industries. The feasibility of drawing up codes of practice for voluntary compliance (such as those the United Egg

Producers' Association has initiated) might also be explored, together with an overall Farm Animal Welfare Council (as in Great Britain) which would advise and coordinate rather than legislate and police. Whether protective legislation is passed should be contingent upon the responsiveness of livestock and poultry producers.

There is an obvious problem with voluntary codes: some farmers will not comply with them if it is to their financial advantage *not* to limit stocking densities. By not complying, they would gain a financial edge over their competitors. Hence, few farmers would be willing to be alone in adopting codes that might lower profits in a highly competitive industry. It is also unlikely, given consumer habits, that many people would be willing to pay more for essentially the same produce just because it came from accredited farms that have adopted welfare codes. One solution would be for producer associations to draw up their own standards and request state or federal assistance to ensure compliance by all farmers. This practice would eliminate any competitive advantage over those conscientious producers who would otherwise voluntarily accept a professional code of husbandry.

There are already some positive signs. Well-designed and well-managed hog-confinement units for breeding and finishing *are* being used in the United States, for example at Ames, Iowa, where no antibiotics are ever given in the feed. With good husbandry practices (including the taking of showers and use of changing rooms by persons entering and exiting the facilities to prevent spread of infection between buildings), the multiple-vaccination programs that are now needed to keep confinement hogs "healthy" can be abandoned. The next great challenge to the veterinarian and agricultural scientist is to develop husbandry systems that do not necessitate or create a dependence upon drugs and vaccines. While the *appropriate* use of drugs and vaccines should not be discounted, thoughtless and habitual use must be deplored. We have long been guilty of this kind of unthinking misapplication of science and technology.

The answer to the problems engendered by drugs and pesticides is in the application of an appropriate (rather than traditional) technology, although this concept runs counter to the compelling arguments made repeatedly by pharmaceutical companies and other believers in the drug cult of modern farming. Farmers have to learn once more how to live with bugs and bacteria, and not feel they have to drug their fields and animals to the point of sterility.

Integrated pest management (IPM) is a new development aimed at reducing the use of pesticides and their cost to the farmer. Heretofore, farmers have used preventive spray programs against pests (both insects and disease), even without evidence of the presence of any pests. The fact is that there are tolerable levels of pests in any field; serious crop losses occur only when those levels are exceeded. Farmers who join IPM pay a fee for an inspector to examine their fields at regular intervals and alert them to any dangerous buildup of pests—only then do they spray. The savings in chemicals and their application more than make up for the cost of the inspector.*

In recent years, two-thirds of American households have changed their diets for health reasons, mainly to cut down on fats and cholesterol. The first casualty was beef. Consumption dropped seventeen percent between 1976 and 1985. Now, the National Livestock and Meat Board is working with the American Heart Association and the American Cancer Institute to identify cuts of beef and pork that meet dietary guidelines for low levels of cholesterol, unsaturated

*Ideally, only farmers practicing organic and IPM agriculture should be entitled to government price supports and other financing by the public.

fat, and sodium. The beef we are getting has become leaner, in response to consumer concerns.

"Welfare veal" and eggs from hens not kept in battery cages are sought after by consumers in some European countries in preference to crate-raised veal and eggs from caged layers. The alternative of raising veal in straw yards has been more profitable under British conditions than crate-rearing, and many consumers and chefs are willing to pay more for eggs from hens kept in aviaries and deep-litter houses. In Holland, free-range and deep-litter eggs cost a few cents more per carton, and producers can't keep up with the public demand. In the United States, there is a significant but virtually unexplored market for lean beef (range- or forage-raised, using less cereal grain).

Coleman Ranches of Saguache, Colorado, is now selling five thousand head a year of "clean beef"—cows who have never been given growth-stimulating antibiotics or hormones, and who eat grasses, hay, and grain that have never been sprayed with insecticide. The feedlot operator and the feeder's veterinarian and nutritionist must sign certified letters that no drugs are given to the cattle. The beef is more expensive, because it takes 20–25 percent longer to raise, but the market (as yet, mostly health-food store chains) is there.[2]

Jim Hightower, Commissioner of Agriculture for the State of Texas, has pledged to move farmers into the marketing stage of the food-delivery system and is organizing direct farmers' markets throughout the state. The economic wisdom of making states more self-sufficient by producing and marketing local produce is slowly gaining momentum across the United States, and state and county organic-farming and regular marketing cooperatives are spreading.*

*For further details, apply to The Cooperative League of the U.S.A., 1828 L Street NW, Washington, D.C. 20036; The Cornucopia Project, Rodale Press, 33 E. Minor Street, Emmaus, Pa. 18049; Rural America,

The passage of the Agricultural Productivity Act (originally the Organic Farming Act of 1981 and 1982) by Congress (it was opposed by the Reagan Administration) gives some indication of political enlightenment. This legislation requires the USDA to analyze and publish studies on twelve farming operations that have used organic farming techniques for at least five years, as well as to establish twelve on-farm research projects to collect and analyze data on the effects of a transition from conventional to organic farming. The USDA is also required to make available material dealing with organic farming and to issue recommendations to all farmers interested in converting to organic farming. The program will cost approximately $2.1 million per year for five years.

Other potentially positive trends include crop rotation to reduce pests and weeds, and hydroponics, and systems using solar energy. But in the animal sector, unfortunately, we have genetic engineering and embryo-transfer techniques to improve the genetic and productivity potentials of livestock, and research into environmental manipulations (such as providing a longer day by using artificial light) to boost productivity.

Hydroponics and aquaculture could be useful tools, but their promise is being compromised by agribusiness practices. Some food processors, such as General Mills, following the pioneering work of small entrepreneurs, are exploring large-scale hydroponic production of vegetables in controlled, soilless environments. These capital-intensive, highly automated systems require little labor, offer year-round production, and will compete effectively with local

1302 18th St. NW, Washington, D.C. 20036. For a national directory of wholesalers of organic produce, write California Agrarian Action, P.O. Box 464, Davis, Calif. 95617, and The Organic Food Production Association of North America (OFPANA) P.O. Box 6414, Lehigh Valley, Pa. 18001.

traditional vegetable farmers and with out-of-season imports from other states and countries. The nutritional quality of such produce is an unanswered question: deficiencies in essential nutrients in otherwise healthy-looking, umblemished produce are a likely problem linked to maximizing profits by lowest-cost nutrient-media formulation.

Another "progressive" development whose benefits and costs are unweighed is genetic engineering. While there may be no immediate economic benefits in genetically engineering giant cows or hogs—or in the creation of chimeras (like the monstrous goat with a sheep's body at the Government Agricultural research station in Cambridge, England, created by fusing together portions of fertilized ova of a sheep and a goat), pharmaceutical companies are already manufacturing growth hormones via this new biotechnology.

The USDA research station in Beltsville, Maryland, has been attempting to create giant sheep and pigs by inserting *human* growth genes into the developing embryos—a technique that has already been used successfully to create giant mice. Such animals may grow twice the size twice as fast, but will require twice as much food. The only saving will be time. Another economic treadmill will be created as superfarms produce animals at a faster rate and thus gain a competitive edge over farmers who raise normal animals.

Spurred by legislation that permits the patenting of seeds and the Supreme Court decision that genetically engineered organisms can be patented, multinational—notably petrochemical and pharmaceutical—corporations have purchased several large seed companies and invested heavily in biotechnology research.[3] They are trying to develop crops resistant to certain herbicides, to create a nice sales package of drugs and seed for future "farmers." It would profit society more if pest-resistant seed stock could be created instead. But then the insect population would shift, and probably new pests would arise as certain species proliferated following the demise of their predators. The research to develop resistant seeds, along with the race to control world seed stocks (germ plasm), could lead to a global monopoly and to the perpetuation of harmful chemical-dependent farming methods.

Other products derived from genetic-engineering technology include new vaccines to prevent diseases and new health- and growth-promoting compounds from industrially grown microorganisms, for use on farm animals.

At this time of writing, some companies are investing in creating new strains of bacteria that may be used in agriculture as pesticides. Indeed, progress has been so rapid that various government regulatory and advisory agencies such as the Environmental Protection Agency, the U.S. Department of Agriculture, and the National Institutes of Health have been caught short and are in a turmoil as to who should regulate what and how. Because of the potentially catastrophic environmental consequences of deliberately releasing genetically engineered organisms into the environment, the urgency of instigating national and international regulations must be recognized; since this new biotechnology is one of the fastest growing industries in the world, corporate responsibility for long-term environmental risks may be overshadowed by the promise of significant short-term profits.

A special report published by the Office of Technology Assessment in March 1985 emphasizes that new technology and genetic-engineering methods will greatly increase the trend towards larger farms over the next fifteen years. The agency concluded that these new technologies will benefit large farms and superfarms, not midsized ones.[4]

Genetic engineering holds promise only if it is integrated into a sound system of food production; otherwise it will be only a temporary palliative. Scientific expertise, capital to invest, and good economic returns are not the basis of a sound agriculture when ethics are lacking. By ethics, I mean the proper *husbandry* of land and stock.

A new set of values, a mental "fix"—not a new technological "fix"—based upon an empathetic as well as rational understanding of and respect for the principles of humane and ecologically sound stewardship, is the antidote to agricide. But the entire system has, to date, been built upon short-term investments, quick profits, and a level of consumer affluence that the system cannot long sustain.

In many European countries, the family farm has been protected for decades by limiting the number of acres a person may own and the size of livestock farms. As a consequence, farms have remained small. With less reliance upon and belief in capital-intensive methods of chemical- and drug-based farming, many are now converting to organic farming. A number of organizations support research and sales promotion, and help with labeling and quality control. The biggest in Europe is Bio-Dynamic Farming. The organization's philosophy is anthroposophy, which emphasizes the unity of the spiritual and the material, and holds that this unity—so necessary for our health and well-being—is in part accomplished through healthful agricultural practices and responsible planetary stewardship. This organization has research centers in Germany, Switzerland, Austria, Sweden, New York and California.

Peter Berg is one of the leaders in the concept of bioregional agriculture, industry, and politics. He sees this idea as a saving formula:

> It's time to develop the political means for directing society toward restoring and maintaining the natural systems that ultimately support all life. *Bioregions* are the natural locales in which everyone lives. *Reinhabitation* of bioregions,

> creating adaptive cultures that follow the unique characteristics of climate, watersheds, soils, land forms, and native plants and animals that define these places, is the appropriate direction for a transition from late industrial society. Environmentalism, at best, reaches its zenith in a standoff. It's time to shift from just saving what's left and begin to assert bioregional programs for reinhabitation.[5]

What will save us is the concern and involvement of the American people. If we refuse to eat food that is too expensive in ecological and humane terms, if we refuse to swallow the chemicals being foisted upon us, if we refuse to accept a staggering national health bill, then the profits that now sustain agribusiness will dry up. American food producers will be forced to provide a product that we will buy. The power of the marketplace is *our* power; it is the only one that agribusiness understands. And we can exercise it to save the earth.

8

Breaking The Cycle of Poverty and Famine: The Role of Humane Sustainable Agriculture

AGRICIDE is not only confined to the United States. It has become a global problem. Agribusiness has contributed to world hunger and poverty by encouraging the adoption of conventional, nonsustainable agricultural practices in developing countries. The so-called "Green Revolution," in which poorer countries were encouraged to accept international bank loans and to purchase hybrid seeds, agricultural chemicals and other agribusiness materials, equipment, and production methods, has caused more harm than good. While some countries like India increased agricultural productivity, primarily for export (and to pay off loans), as a consequence of the "Green Revolution," rural poverty and hunger have actually worsened. The primary beneficiaries of this revolution are evidently the banks, brokers, and government advocates of industrial agriculture, along with the multinational agribusiness corporations whom they serve.

Whenever the cycle of poverty remains unbroken from one generation to the next, environmental degradation inevitably intensities. Whenever there is poverty, family income is dependent upon child labor, especially to help gather firewood, work in the fields, and tend grazing livestock. So poor families have many children. But with more mouths to feed, the downward spiral toward environmental destruction and famine becomes inescapable. It is a tragic irony that increasing poverty correlates with increasing population.

> Where the amount of land allocated is based on the ability to cultivate it, this ability—under the low-resource farming conditions prevailing in most of sub-saharan Africa—is primarily determined by the ability to mobilize labor. In most cases, this means family labor—more specifically, female and child labor. Indeed, a number of field studies report this to be an important incentive to increase family size through such means as polygamy and/or pressure on women to have many children.
>
> The controversial suggestion derived from the above is that there is high economic value to rural Africans of having large numbers of children. Larger families appear to do better economically than small families. Children contribute labor in cropping, livestock herding, fetching water and fuelwood, and child rearing. As forests, water availability and soil fertility decline, farmers and pastoralists obtain less product per hectare. The only resource available to them to increase production is family labor and increased farm size. Hence, agricultural stagnation and environmental degradation, in resource-poor situations characteristic of sub-saharan Africa, provide an economic incentive—and often a survival strategy—to maintain large families."[1]

Poor families and communities that traditionally keep livestock and who do not have good arable land to grow their own food cause great harm to the environment, especially in arid and semi-arid savannas, grasslands, and bush country. That some cultures, like that of the Australian aboriginal peoples, have kept their numbers well within the carrying capacity of the environment for thousands of years, attests to the vulnerability of agrarian and pastoralist societies compared to the frugal efficiencies of a more sustainable gatherer-hunter way of life. This way of life necessitated small nomadic family groups with few possessions to encumber mobility. It imposed strict limits on population growth in relation to the limits of available natural resources.

No downward spiral of declining soil fertility and dwindling natural biodiversity was triggered until the advent of the plough and pastoralism. Developments in agricultural technology and the green revolution in Asia have temporarily arrested the former downward spiral. Science and technology, as Shridath Ramphal concludes, have not prevented us from borrowing the present from the future.[2] They have helped enlarge our options, and postponed our arrival at that point where we exceed the Earth's carrying capacity.

In the interim, environmental degradation and the irreversible loss of such vital natural resources as topsoil and fertile arable land continues apace. Poorer countries also lack the resources to restore vast areas of degraded agricultural and range land, and the downward spiral continues. A vicious circle of more and stronger pesticides, chemical fertilizers, genetically engineered products from new-generation livestock vaccines, crop sprays, and seed stock; of more aid and "development" programs, etc., follows, at great cost to all. Technological innovations, no matter how good the scientific rationale, don't work if destructively nonsustainable agricultural and animal husbandry practices aren't changed. New technological inputs like chemical fertilizers, pesticides, cultivators and harvesting machinery, genetically engineered seeds, and livestock vaccines may temporarily enhance productivity. But the agricultural systems to which these technological "fixes" are applied in the first place must be inherently sustainable and ecologically sound, as well as culturally acceptable and socially just. Regrettably, in the absence of inherent long-term sustainability, such inputs create costly dependence, almost invariably aggravate environmental deterioration, and lead to a loss of indigenous knowledge. The possibility of environmental restoration and the establishment of programs to improve soil fertility and rangeland productivity then become ever more remote.

In a review article on the topic of animals in Third World agriculture, Ward et al.[3] make the following assertions:

> Production of most livestock in the developing world is far below its potential. For example, the annual 'offtake rate' (yield) for cattle herds in Africa averages about 3 percent compared with some 35 to 40 percent in North America; milk production in Southeast Asia is estimated at 37 kilograms compared with 480 kg in North America (where over 80 percent of the cows are of beef breeds with relatively low lactation rates) and 1289 kg in Western Europe (which has a much higher percentage of dairy cows and fewer specialized beef cattle), and 1373 kg in Japan. The principal constraint is the quantity and quality of feed . . . In many cases feeds of poor quality will not even support maintenance, and as a result body reserves are used until more favorable climatic conditions recur, at which time reserves can be replaced . . . Because of the escalating demands on the land for human food, livestock frequently depend upon scavenging what people cannot or will not eat. Even where land is available for forage production, its quality tends to be low because the focus is on food and cash crops . . . Next to inadequacies of feed, cultural patterns

> are the most serious obstacles to livestock development. Because of the multiple uses of livestock, many of them unrelated to productivity, numbers are prized over economic productivity. Owners keep large numbers of animals which they are unable to feed adequately, and *the condition of many animals is little short of what would be considered criminal* in North America. Animals in poor condition are prime targets for parasites and disease, and despite the vast strides made in veterinary care the health status of millions of animals remains alarmingly low. [italics mine]

The veterinary profession in the developing world will do more harm than good if its focus on livestock and poultry health, and productivity, is limited to dispensing costly vaccines and various pharmaceuticals, while there is no sound program to improve animal nutrition and husbandry practices.

The nonmedicinal prevention of livestock disease in African rangeland ecosystems has been discussed by Carles,[4] who identifies low-input correctives within the ecosystem itself and changes in husbandry practices that improve animal welfare and reduce the spread of disease. He concedes that costly medicinal interventions are sometimes ineffective and deleterious to the environment. Some of his suggestions include the following: providing shade and shelter; keeping young stock away from contaminated night enclosures, watering points, and mineral licks; ensuring hygienic disposal of infected carcasses; developing breeding programs to optimize productive potential of indigenous genotypes; reserving good rangeland for young and pregnant animals; relocating night enclosures every 7–10 days to reduce infections; and avoiding contact with other flocks and herds, especially at watering points.

Some two-thirds of the world's agricultural land is in the form of permanent pasture, meadow, and range, and about 60 percent of this is not suitable for cultivation.[5] Much of this land is severely degraded. Husbandry practices like rotational grazing and mixed species grazing, along with appropriate reseeding with forage, nitrogen-fixing trees, and perennial plants, and the judicious use of fertilizers where soils are nutrient deficient, are urgently needed. Poor livestock nutrition means poor productivity and poor health.

Livestock nutrition can also be improved by better livestock/crop integration. In China, Mao Tse Tung encouraged pork production with the assertion that "every pig is a fertilizer factory." Livestock (and also poultry and fish) play a vital role in such nutrient cycling and in converting crop residues and food byproducts into animal protein and fiber.[6] Better integration of livestock and cropping practices can bring many advantages other than improved livestock and poultry nutrition and health.[7] These include soil improvement via recycled animal wastes and crop/forage rotations, which also help reduce soil erosion on arable land. Such practices also help reduce the need for herbicides and pesticides since chickens, geese, goats, and other livestock are natural controllers of crop pests and various weeds.

Ward et al.[3] conclude that:

> To make an animal system possible, the nutritional constraint in the tropics requires for the Third World the equivalent of the turnips and clover that revolutionized agriculture in Flanders and England. Given the finite supply of land, the solution lies in more intensive use of existing arable land. Some prototypes exist: cassava is a high-yielding plant under cultivation in the tropics, and sugar cane is being used as a basis for animal economies in several tropical countries (6) . . . To provide the needed supplementation as well as improve soil fertility, multiple cropping can be applied to produce forages, especially legumes which fix nitrogen. Other crops now being used for this purpose include berseem in Egypt,

> and *Stylosanthes* grasses in tropical Africa, introduced after successful experiences in Australia . . . The entire question of multiple cropping needs renewed emphasis in development programs, especially with crops of use as animal feeds. Techniques in interrow planting, for example, underplanting of forage crops among tree crops, have been successfully employed in several areas of the world; pastures too can be established among tree crops where livestock are available to graze them . . . Increases of over 100 percent in total production have been reported when maize was intercropped with cowpeas, compared to cowpeas raised alone, with any of four different methods of tillage, including zero tillage. Surprisingly, zero tillage gave the highest yield of the four.

Increasing human and livestock populations, coupled with nonpolluting tsetse fly control and multiplication of trypano-tolerant cattle breeds like the N'dama, will seriously endanger the resource base over the next decade in sub-saharan Africa. Hence, there is an urgent need to intensify and diversify crop production in adjacent higher rainfall areas, increase the use of crop residues and by-products, and also plant more fodder trees, like *Leucaena*, to augment the productivity of these pastoral systems. (The utilization of the latter can be enhanced via holistic resource management, which includes carefully monitored, short-duration, high intensity rotational grazing.)

Advances in feed technology, such as treating straw with urea or ammonia or feeding straw with urea and molasses, can also be utilized, especially for dairy cows.

Greater emphasis is needed on genetic improvement of crops in the direction of improved nutrient value of crop residues for livestock, a direction overlooked by the green revolution. The lack of crop/livestock integration, like the lack of forage seed production and the adoption of grass/legume rotations, can only be corrected when the veterinary livestock service is more closely integrated with crop extension service. A unified extension service is clearly needed.

Developing nations should focus on low-cost/low-input ways of improving livestock and poultry production, such as by utilizing homegrown cassava and sweet potato meal as a substitute for imported feed grains. The dietary protein needs of pigs and poultry will continue to compete with human needs. But advances in biotechnology may soon enable essential amino acids like lysine, methionine, and tryptophan to be synthesized from biomass sugars at low cost for incorporation in animal feed. Other developments in biotechnology, like porcine and bovine growth hormone (rBGH) to increase pork and milk production, are high cost inputs that have no place in a humane sustainable agriculture.

Farmers around the world have traditionally adopted mixed cropping and integrated livestock farming practices in order to maximize the use of scarce resources, and to minimize disease or drought that could wipe out a single crop. In sum, they evolved methods of farming that helped to cope with natural disasters by maintaining diversity in both genetic seed stocks and in the kinds of crops and livestock they raised.

The urgent need to prevent this cornerstone of sustainable agriculture from being permanently displaced by hybrid monocrop farming (which usually necessitates high inputs of chemical pesticides, fertilizers, etc.) is not widely recognized by Third World governments. Preserving diversity means protecting traditional farming practices which, according to several studies, are more sustainable than conventional monocrop industrial agriculture. In many countries this will entail land reform and programs to help protect native seeds.

The Bush administration's refusal to sign on to the 1992 international treaty to protect biodiversity—plant and animal species and varieties—was aimed at protecting U.S. biotechnology interests. This industry wants no constraints on the patenting of new products derived from the genetic resources of the forests and grasslands of the Third World which might generate great profits in such areas as new cures for cancer or disease resistance in corn. Third World countries would want royalty payments on these biotechnology products which utilize genetic material from their own natural resources. The biotechnology industry, in claiming that these resources are the common heritage of humankind, are using rhetoric to justify the continued exploitation of poorer countries. One case in point is the United Kingdom's Institute of Tropical Agriculture, which provided funds to a British university to isolate the gene mechanism that made Nigerian cowpeas resistant to weevils. Nigerian farmers who had spent decades selectively breeding these resistant seeds, were cut out of any royalty payments when a biotechnology company took out a patent on the identified gene and began licensing seed companies to incorporate it into a variety of different crops.

In discussing this and related examples of colonial exploitation, and of technology transfer that could cause social chaos and environmental ruin to Third World countries, analyst Fred Pierce[9] concludes:

> The local development of products and processes tailor-made for local conditions, and the exchange of ideas between countries at similar stages of development, is a more promising model for successful technology transfer than blindly importing alien Western technologies. Looked at this way, indigenous knowledge is at least as valuable to Third World countries as Western scientific skills. The trick is to marry the two.

Clearly there is nothing intrinsically wrong with science and technology. But when technology is applied within a narrow paradigm of productivity or emergency relief, or is primarily utilized to enhance corporate profits, the nemesis of agriculture is inevitable. Agritechnologies, including advances in veterinary medicine and biotechnology, must be applied within a far more holistic, organic framework so that such innovations and inputs are not short-term band-aid remedies, but are integrated with a long-term program and commitment to agricultural sustainability.

Third World countries have yet to fully grasp the fact that in spite of its wealth, science, and technology, the Industrialized World does not have a sustainable agriculture. Third World countries should not forget how totally destructive the Industrialized World's colonial agricultural developments were throughout the 19th and 20th centuries to their own often highly sophisticated and sustainable agricultural traditions. And Third World countries should not be seduced or coerced into accepting agricultural programs and veterinary and agritechnologies from the Industrialized World that do not accord with the principles and practices of humane sustainable agriculture. (See Addendum at the end of this chapter for Principles of Humane Sustainable Agriculture.) Mahatma Gandhi, when asked if India, upon its independence from British exploitation, would ever attain British standards of living, said, "It took Britain half the resources of this planet to achieve its prosperity. How many planets will a country like India require?"[10]

How many will Africa need? Africa could lead the world into the 21st century, given the right aid and sharing the right vision, not of material prosperity and industrial productivity, but of material security and self-sustaining industry, especially in the agricultural, forestry, and mining sectors. If Africa uses the same technologies from the West, and imports the same consumptive values, then its poverty, hunger, and disease, along with increasing social injustice, urban crime, violence, and war, will intensify. This is happening in the Middle East and Eastern Europe today, where environmental degradation and industrial inefficiencies and pollution, respectively, have contributed to economic collapse and internecine strife. And the increasingly dysfunctional economies of the industrial democracies of the world mirror the increasingly dysfunctional condition of consumer-oriented cultures that are still the role model that many developing countries seek to emulate.

World population is predicted—the AIDS epidemic notwithstanding—to reach 15 billion by the year 2100. Agricultural economist, Jonathan M. Harris, asks how all these people will be fed when there are presently 2 billion people living in poverty and 1.6 billion malnourished in the world today.[11]

An estimated 2 billion hectares have been abandoned as a result of agricultural misuse during the history of agriculture.[12] Harris points out that most of the 1.5 billion hectares of arable land that are available worldwide are already under intensive production, along with millions of hectares of fragile marginal land.

With soil erosion continuing worldwide, the land available for agriculture cannot sustain any further increase in the human population. Nor can it sustain the current population of 5.4 billion. Costly inputs of chemical fertilizers are not affordable to the most needy. One third of the fossil energy agricultural uses is for fertilizers in the United States, where, according to Harris, soil is being lost 18 times faster than it reforms, and some 35 percent of its arable land has been abandoned because of soil degradation. Efforts to increase the productivity of the land by large-scale, mechanical cultivation; continuous cropping; the ploughing of marginal, erodible land; and the clearing of forest and grassland cover—all in the name of profit and need—continue to limit the capacity of the Earth to sustain us and the generations to come.

The costly use of pesticides to boost crop production is another factor contributing to the failure of conventional agriculture. Some 2.5 million metric tons of pesticides are applied annually worldwide to protect crops, yet still about 35 percent of all crops are lost to pests and a further 20 percent to other pests after harvest.[13] Over 500 species of insects, 270 species of weeds, and 150 species of plant pathogens are now resistant to pesticides.[14] And, according to the World Health Organization, 1 million people are poisoned and 20,000 killed by pesticides each year. Also, uncounted billions of beneficial insects and birds, as well as other wildlife, both terrestrial and aquatic, are harmed by these agri-poisons.

Yet another soil problem is waterlogging and salinization, occuring often as a result of costly and wasteful crop irrigation. Harris estimates that globally 80 percent of fresh water is used for irrigation purposes. Deep aquifers are drying up, as are many lakes and rivers, many of which have been polluted with agrichemicals. The construction of hydroelectric dams have caused further, irreparable ecological damage and loss of good agricultural land.

Harris predicts a collapse of agricultural production well before 2050, "if not on a global scale, at least in large regions."[11] He recommends a reduction in the number of people on Earth to a sustainable population of around 1 billion, coupled with soil, energy, water, and biological resource conservation and adoption of ecologically sound and sustainable agriculture.

Harris is not a prophet of doom. He is a realist. We must all face reality and not use denial or false hope in science or some technological fix, new vaccine, or supercorn to avoid or delay constructive and concerted intervention.

We must also be wary of our misguided altruism which so often prolongs suffering, for example by supplying food relief but doing nothing to facilitate environmental and cultural restoration, such as adopting an ecologically sound and sustainable agriculture, coupled with family planning. And we must all work together, for we are but one people of one Earth.

Our agriculture and all our earth-dependent and exploiting industries will continue to be dysfunctional and cause ever more harm than good until the principles of a humane and sustainable global community are put into practice worldwide. Our country is, indeed, this planet.

Since agriculture is the economic cornerstone of national security, every nation, the developed and less developed alike, should begin the task of saving human civilization from further suffering and chaos by making agriculture socially just, humane, and sustainable.

Winrock International's 1992 *Assessment of Animal Agriculture in Sub-Saharan Africa*[15] underscores this need to feed Africa's burgeoning urban populace on a sustainable basis. The report states, "Crops and livestock can no longer be viewed as separate and inevitably competitive enterprises," because both crops and livestock "have essential and interconnected roles to play in the future development of agriculture in sub-Saharan Africa."

The report further states, "If greater agricultural sustainability is to be achieved and if adverse environmental effects of cultivation are to be minimized, livestock must be properly utilized in agricultural development processes . . . A significant increase in productivity of the rangelands in the arid zone and the dryer portions of the semi-arid zone is not economically feasible."

Winrock International's conclusion is that, since "Urbanization will force the commercialization of agriculture and increase the demand for foods of animal origin," then "An expansion of intensive commercial production of poultry, pigs and dairy is envisioned." This, however, can be challenged. Is this conclusion a prediction or an advocation? The urban demand for foods of animal origin in the industrial world has spurred the development of a nonsustainable animal agriculture. Why would Africa be any different?Dairy products and a little veal, beef, and chicken may be produced sustainably, but the "intensive, commercial production of poultry and pigs" may be antithetical to the long-term good of Africa. Corn and other high-energy and protein crops, better fed to people, will be needed for these poultry and pigs, along with a host of other costly inputs to control diseases. If Winrock's panel of experts had been more culturally diverse in view and wisdom, they would have been more sensitive, perhaps, to the fact that one-third of the African continent is Muslim[16] and does not regard the pig as food. If an Asian advisor had been on Winrock's panel, they might have suggested the intensive commercial production of dogs, almost as omnivorous as pigs, and highly profitable, as is done in Korea and Vietnam.

But more to the point. This report considers the continuing urbanization in Africa a given, a trend from the developed world's perspective that should be strongly discouraged. By focusing on the present and future needs of urban consumers, the needs, interests, and rights of traditional sustainable village communities are neglected. Yet it is these communities, if provided the right infrastructure, that could be the sustainable food-source for urban peoples of the future. They should not be sacrificed in the name of progress, but should be integrated and their farming practices improved as needed, not changed to provide feed for factory-raised pigs and chickens. This has been the undoing of the family farm and sustainable agriculture in the United States and Europe.

Winrock's report is not without merit, however. Even though it raises the controversial specter of safari hunting and the harvesting of wildlife, it acknowledges that most livestock development projects in Africa have focused on "efforts to increase the productivity of the rangelands and the off-take of animals in pastoral production systems. Most of the latter projects were based upon an insufficient understanding of African rangelands and the pastoral systems that utilize them. Thus many of the projects were improperly directed." The tragic legacy of pastoral and rangeland development projects in Botswana supports this conclusion. The mission-oriented focus of the veterinarian and animal scientist to reduce livestock and poultry diseases and to increase the productive utility of these animals is based on objective science. But it is not always based on sound reasoning or economics, as when intensive husbandry systems cause animal suffering and disease, harm the environment, and even jeopardize public health and the livelihood of local farmers. The science of animal production must be holistic and not so mission-oriented that it is not integrated with other disciplines such as steady-state economics, ecological science, bioethics, rural sociology, and cultural anthropology.

No matter how good the sciences of animal production and disease control may be, they will do more harm than good if they are not adequately funded. Without adequate funding to control insect-borne cattle diseases in Africa in the ecologically least harmful way, but instead with funds just sufficient to prevent epidemics of foot and mouth disease and rinderpest, there is no incentive to reduce herd size. This is because individual productivity is low when large numbers of livestock are chronically ill, so large herds of poor vitality become the norm. This problem is compounded by the cultural tradition of valuing livestock herds and flocks on the basis of how many head there are rather than upon objective measures of individual productivity.

In turn, this problem, coupled with competing interests over the rights to traditional grazing lands, means that Africa's livestock industry will create more deserts and strife if it is not radically and effectively reformed in accordance with the principles and practice of humane sustainable agriculture. Concern over the worldwide, near-extinction of predator species, like the wolf in Europe and North America, the wild dogs, leopards and cheetahs of Africa and Asia is a concern that is antithetical to the historical tradition of the ever-expanding livestock cult and industry that continues to sacrifice biological diversity on the grounds of custom and economic necessity.

No further expansion of the world's livestock industry should be encouraged until the existing population becomes healthier, more productive, and also less destructive and wasteful of rangelands and other natural resources. In many regions of the world that are afflicted by cycles of drought, intensive rangeland restoration programs, fodder banks, and improved veterinary services (from vaccination to artificial insemination) are urgently needed. Further encroachment by pastoralists and cattle ranchers into all remaining wildlife regions should be strictly prohibited. Socially just, equitable, and ecologically sound agricultural activities that benefit indigenous peoples first and foremost, should take priority over other agricultural practices that are ecologically unsound, export oriented, and may or may not be linked with the disenfranchisement of indigenous farming communities.

The opening up of wildlife preserves for trophy hunting is ethically questionable. It should be strictly regulated from an ecological and humane perspective to maximize natural biodiversity and to provide significant material benefit to local peoples.

Nomadic and semi-nomadic pastoralists, like the Masai, and migratory species, like the elephant and wildebeest, pose special management problems when their normal migratory travel routes and traditional seasonal rangelands have been taken over for more intensive livestock production and for other purposes such as the operation of plantations and the mining of diamonds, gold, and other natural resources.

Following a detailed ecological and demographic inventory, Tanzania, like other sub-saharan countries such as Zaire, Botswana, Kenya, and Zimbabwe, should be able to protect what natural biological and cultural diversity remains within their democratic jurisdiction. And, with appropriate aid from donor nations of the industrial world, it should be able to establish core wildlife sanctuaries, designated as World Heritage Sites to protect them from further encroachment, fragmentation and disintegration. Buffer zones that permit some harvesting of natural resources, intergrazing of livestock and wildlife, and protected corridors for both migratory species and nomadic pastoralists, will require dedicated commitments and challenge the ability of these nation states to work cooperatively, since their boundaries are artificial and political rather than biological and cultural. Efforts to preserve the natural and cultural diversity of this sub-saharan region could be the catalyst for a truly democratic union of post-colonial African nations. But so long as they continue to experience conflict internally and across their political borders, the demise of all that is natural and beautiful in Africa will be assured. The consequences to the world will be profound, not the least of which will be the impact on the global climate as deserts spread and the great rainforests, like those of Zaire, Zimbabwe, and Tanzania, disappear forever. The plight of Ethiopia and the Sahel will be the plight of all the semi-arid countries south of the Saharan sands if cooperative and compassionate solutions are not quickly implemented.

What is the primary justification for the continuation of inhumane, unethical, and nonsustainable livestock and poultry production practices? The goals of progress and profit too often mask the realities of ignorance, indifference, ideological rigidity, arrogance, and greed. No less disturbing are the rationalizations used to justify cruel and nonsustainable livestock husbandry systems, and the blatant denial of the concerns for the legitimacy of farm animal welfare and of environmental protection advocates that are dismissed as anti-establishment and anti-progress.

Four examples of such concerns will suffice:

1. The huge feedlots of 200,000 to 500,000 cattle in the United States and Australia and the vast confinement factories raising 500,000 to 1 million pigs and poultry in the United States squander grain to fatten animals for human consumption and create a costly and environmentally hazardous waste management problem. These intensive systems are heavily subsidized and are relatively overproductive, which puts small, traditional farms out of business and harms third-world farmers when exported produce is "dumped" at below fair market price.

2. The U.S. government continues to use cruel and indiscriminate methods of predator control to protect ranchers' stock, including gassing, trapping, poisoning, and burning animals in their dens. The predator control program cost taxpayers $29.4 million in 1990, which some analysts contend is more than the amount of losses caused by predators. In 1989, the U.S. Department of Agriculture's Animal Damage Control Program killed some 7,158 foxes, 86,502 coyotes, 236 black bears, and 1,220 bobcats, along with thousands of other species.

3. The functional integrity of the global atmospheric ecosystem is being disrupted by the massive burning and clearing of rainforests in Central and South America to raise cattle for beef export.

4. The World Bank- and European Economic Community-subsidized cattle industry in Botswana has helped impoverish and disenfranchise thousands of small farmers and has severely

> undermined traditional, more sustainable agricultural practices. This is a classic example of how ill-conceived aid and development programs benefit so few and cause more harm than good. Thousands of miles of cordon fences to control foot and mouth disease in Botswana have meant the demise of millions of migratory species, like the wildebeest and zebra, once a sustainable resource for thousands of traditional African societies. They are a tragic monument to the world of the destructive consequences of a nonsustainable and socially unjust livestock development project.

More and more people are beginning to realize that their own fate is inevitably linked with the fate of endangered and threatened species, and that the protection of such species is part of the solution to the accelerating deterioration of the global ecosystem and the social, economic and spiritual disintegration of industrial and developing nations alike. We do not seem to know when to say "enough," for our needs and wants continue to multiply beyond the carrying capacity of the Earth. Our most precious and selfish wants—more children and grandchildren, more livestock and land—will be our undoing, because this planet cannot contain what it cannot sustain.

We are as far removed from our gatherer-hunter ancestors as we are disconnected in our unnatural lives from the ancient roots of natural farming. There was a time when we were all acutely aware of our dependence upon nature. But now we are dependent upon the food-mart, the power companies, the gas station, and municipal sewage, water, and garbage removal services for which we pay an inordinate environmental cost. The real cost of living includes a hidden cost that we would see as outrageous, even suicidal, if we were not so disconnected from nature, from the land, the ultimate source of our sustenance and being.

It is a cost the Earth cannot sustain because those of

us caught in the all-consuming web of industrial society are not living in a sustainable way. We enrich the agribusiness and the petrochemical and power companies which in turn deplete and poison the Earth. This mummifying web is becoming ever larger as the rainforests of the world are cut down, wetlands drained, rivers dammed, and the oceans poisoned and plundered of life.

The once seamless web of life, the planetary ecosystem, is now in shreds. We have been slow in learning that we cannot progress for long at nature's expense without respect and reverence for the Earth and all Creation. We are realizing, too late some fear, that we can only progress when our agricultural and other essential industrial enterprises are developed as analogs of natural systems in that they are self-sustaining and contribute to the functional integrity of the biosphere and to the life community as a whole.

Ecologists are now researching how small an area of forest or swamp can be in order to maintain its optimal level of biodiversity. And agronomists are discovering how a monoculture of rice, wheat, cabbages and corn must be managed in order to remain healthy, resist invasion by "weeds," insects, fungus molds, and virus blights, and reduce the need for synthetic fertilizers and costly pesticides.

The next step is for ecologists and agronomists to work together to restore and preserve natural biodiversity and enhance agricultural sustainability and long-term economic and agrocological stability.

Ecological or natural farming is one way that we can begin to repair the web of life. We will begin to free ourselves from the all-consuming web of a nonsustainable, consumptive way of life when we, as urban consumers, support natural farming. We must selectively boycott the food-marts and help ensure that local and regional produce from natural farms fulfill all our basic dietary needs. These diverse, regionally optimal organic farming systems will be based upon the principles and philosophy of humane sustainable agriculture. Their establishment is as urgent in Third World countries as it is throughout the developed Industrialized World.

Addendum: Humane Sustainable Agriculture Program

American agriculture's gains in production and efficiency have long been widely heralded, often in the context of feeding the world. A host of economic, environmental, and social problems have accompanied these gains, however, raising questions about whether these indirect or hidden costs are justified and whether production can be sustained.

Chemically treated and fossil-fuel dependent crop monocultures and high-density livestock and poultry facilities pollute the environment, including drinking water, impoverish the soil, and jeopardize farm animal health and well-being. Once thriving rural communities suffer declines in number of family farms, jobs, and population. And hunger and malnutrition, both at home and abroad, have increased.

Humane Sustainable Agriculture (HSA) addresses these problems by analyzing costs and benefits of crop and animal production more broadly than does conventional agriculture. This means looking at costs and benefits that are public as well as private, long-term as well as immediate, and those that can be given monetary value as well as those that cannot.

The Humane Society of the United States (HSUS) HSA Program promotes:

/ farming and ranching with compassion and consideration for land and animals;

/ joining together productivity and efficiency with the related goals of better husbandry of soil, crops and livestock, and careful stewardship of natural resources;

/ understanding that livestock and poultry have an integral role in alternative, sustainable agriculture; and,

/ raising livestock and poultry under conditions that satisfy their physical and behavioral needs, just as the land is treated in the ecologically sound manner of good stewardship.

The HSUS encourages farmers who espouse humane, sustainable practices, avoiding dependence upon antibiotics, hormones, pesticides, herbicides, synthetic fertilizers, and inappropriate biotechnologies. At the same time, the Society urges consumers to see the importance of their food-buying decisions.

HSA offers society a way to "eat with a conscience"—a way to support compassionate stewardship of farm animals and the environment. (see p. 213 for details)

In sum, HSA is a life-enhancing food production system that embraces ecological and humane principles, the goal of environmental and consumer health, and the ethics of social justice, which includes respect for all nations' desire for food security.

9

Genetic Engineering and Our Farming Future

VENTURE capitalists are investing heavily in a new biotechnology that could rapidly alter agriculture even more, turning crops and farm animals into pharmaceutical factories and making natural food a thing of the past. By deleting certain genes and inserting the genes of other species, gene-jockeys in private laboratories and corporate-funded universities are developing a whole new generation of life forms. This Second Creation of biocommodities promises to obliterate much of the First Creation or the natural world, including natural biodiversity, ecologically sound farming systems, and the wholesome products thereof.

What's left of the countryside will soon be transformed into a bioindustrialized wasteland of commodity crops that produce various drugs, vaccines, industrial oils, biofuels, and biochemicals, as well as genetically engineered food ingredients, and feedstuffs for genetically engineered animals. These animals will be incarcerated in vast factories to provide us with not only meat, eggs, and dairy products, but also with pharmaceuticals, human blood, and organ parts. Already, pigs have been engineered to produce human blood and to have human immune systems so they can serve as donors for people in need of a new heart, pancreas, or liver.

Throw away the sci-fi tabloid images and predictions of monster plants and pigs the size of elephants emerging from the corporate cornucopia of this new biotechnology. The monster that is emerging is no genetically engineered mutant life form. Rather, it is a monster of a different order that seeks to monopolize and manipulate the genes of life for profit and to commoditize (under patent protection) whatever it can create and market. What's behind this "bioexplosion" in investment, research, and development in agricultural biotechnology is not the altruistic goal of feeding a hungry world, or creating a humane, socially just and sustainable (ecologically sound) agriculture. Rather, profits and monopolistic control over the world's genetic resources dominate this "bio explosion." New varieties of genetically engineered seeds, livestock, and poultry will be patent-protected as intellectual property. Patent protection doesn't seem all that much of a problem on the surface, but it is when you realize that the multinational pharmaceutical and petrochemical corporations and mega-agribusiness commodity brokers (the global grain merchants) have been buying up seed companies worldwide over the past decade. Farmers will soon have access only to patented, genetically engineered seeds, like corn and wheat, that have been engineered to be resistant to the corporation's own brand of herbicide and to genetically engineered varieties of livestock and poultry that they will raise as contract growers under corporate peonage.

Corporate gene-jockeys are also creating agricultural crops that produce their own insecticides and fertilizers, although it would be cheaper and better for farmers to adopt such tried and true traditions as crop rotation to control pests and enrich the soil. But the agribusiness soil-mining industry doesn't see such ecologically sound practices as progressive or profitable. They dismiss the fact that genetically engineered traits such as herbicide resistance, insect resistance, stress tolerance, and nitrogen-fixing ability could be transferred from crops to weeds through hybridization to produce new and more troublesome weeds and insect pests.

The U.S. Department of Agriculture (USDA) now allows the release of genetically engineered crops under the old Federal Plant Pest Act. This is a wholly inappropriate statute that was drafted to control agricultural pests, not genetically engineered crops. The USDA is now proposing to eliminate federal permits for field testing in order to speed bioengineered products to the market. All that must be done is to notify the USDA of the intended field test.

In spite of potential risks, by 1994 the U.S. government had approved over one thousand deliberate test releases of genetically engineered organisms into the environment under the flag of scientific "field trials." All to what end, one might wonder, in view of the chronic overproduction of crops in the United States most years. Some biotechnological companies say these new generation crops will reduce the need for harmful chemical pesticides and fertilizers (which a decade ago we were told were safe). The Environmental Protection Agency (EPA) has the task of regulating the release of these bacterial pesticides under the same rules that were developed to ensure the "safe" use of chemical pesticides, which is absurd. Chemical pesticides can't multiply like harmful bacteria.

Under the Bush administration, the President's Council on Competitiveness, headed by Vice President Dan Quayle, sought to undermine proposals to improve the regulatory scope of the EPA and USDA on behalf of the U.S. biotechnology industry. This was a counterproductive move, since such deregulation only fanned the flames of public concern over the safe and appropriate use of this new biotechnology in agriculture.

The biotechnology industry claims that it will contribute to agricultural sustainability. But what is sustainable (or ecologically sound) about developing herbicide resistant crops? Genetically engineering crops to be more nutritious and to contain more amino acids and trace minerals isn't necessarily a good idea either when soil quality is ignored and the soil-building and enriching attributes of sustainable agriculture are short-cut. There may be shortcuts to making money, but there are no shortcuts to achieving a sustainable agricultural system.

In my recent book *Superpigs and Wondercorn*,[1] the inevitable conclusion is drawn that genetic engineering, in agriculture, is being applied as a band-aid to prop up a nonsustainable system that has led to the demise of 450,000 family farms in the United States since 1985. This nonsustainability is due to extensive soil erosion and surface and groundwater pollution, in part, to monocrop agriculture (corn, soya beans, and other agricultural commodities for export and to feed livestock and poultry); and to increasingly inhumane, "factory" farming systems of livestock and poultry production. I was unable to find any evidence that agricultural biotechnology was being developed in a coordinated way to improve the quality of rural life, the quality of food, the quality of the environments, or the well-being of farm animals (except for new-generation vaccines and diagnostic kits).

As for food quality, it is not farfetched to envisage a menu that includes a variety of genetically engineered products that have been redesigned to resist frost (like tomatoes and strawberries containing fish antifreeze genes) or to have a longer shelf-life, which means they won't look bad even if essential nutrients have long deteriorated.

Regardless of the fact that consumers may be exposed to a host of new and unsuspected allergens in genetically engineered foods, the Food and Drug Administration (FDA) has, to date, refused to consider labeling genetically engineered foods. This same agency has bent over backwards to

A DINNER OF TRANSGENIC FOODS

Appetizers
Spiced Potatoes with Waxmoth Genes
Juice of Tomatoes with Flounder Gene

Entree
Blackened Catfish with Trout Gene
Scalloped Potatoes with Chicken Gene
Cornbread with Firefly Gene

Dessert
Rice Pudding with Pea Gene

Beverage
Milk from Bovine Growth Hormone (rBGH)-
Supplemented Cows

Federal permits for environmental release are pending or have been granted for all the transgenic plants and animals included on the menu. rBGH is under consideration for approval as a veterinary drug.

Source: *The Gene Exchange*, December 1991.

get the new genetically engineered bovine growth hormone r (rBGH) approved for injection into dairy cows (to boost milk production) in spite of unresolved consumer health and animal welfare concerns and chronic overproduction in this agricultural sector.

Genetically engineered rBGH — which is an affront to the science and ethics of good dairy cow husbandry — is the first product that the biotechnology industry has yet to recognize as their own Ford Edsel. And selling it to good dairy farmers is like convincing Eskimos that they need refrigerators.

Without a strong alliance between farmers, ranchers, and consumers to preserve the natural food market and to help establish a humane sustainable agriculture, this new technology may well transform the food industry into an enterprise that is likely to lead to "fieldless foods." A wide range of genetically engineered food products and ingredients could be manufactured from raw biomass materials like corn or tree-extract sugars, and others grown in culture vats using the gene-based elements of cell or tissue cultures from various fruits, vegetables, and animal parts.

Even if the "fieldless foods" scenario isn't quite on the horizon yet, we should be mindful that the direction being taken by agricultural biotechnology is the *opposite* of that being urged by advocates of organic and other alternative sustainable agricultural practices. Using rotational grazing, for example, dairy farmers have been shown to get more milk from healthier and happier cows than from those injected with rBGH and fed concentrates in confinement sheds.

Profound structural changes are now taking place in agriculture and the related food industry as multinational corporations gain a monopolistic control over how and what farmers farm and what kinds of food people eat. Agri-biotechnology monitors Cary Fowler et al put it this way:

> What is now emerging throughout the corporate sector in the United States, Europe and Japan is a new, unprecedented institution of economic and political power; the multifaceted transnational life sciences conglomerate—a huge company that will use genes to fashion life-necessary products just as earlier corporate powers used land, minerals, or oil.[2]

The world's food system could become extremely vulnerable in many unforeseen ways if developments in agricultural biotechnology are not based upon the ethics and principles of humane, sustainable, and socially just agriculture. Jack Doyle, in his book *Altered Harvest*, expresses this concern:

> Today, we may be moving toward a high-tech, house-of-cards agriculture worldwide, with genetic engineering at its base; a system in which one monkey wrench or one unforeseen mutation can create enormous problems. Just as the technology of hybrid corn production went wrong in 1970, aiding the advance of the corn blight, the agricultural biotechnologies of genes, microbes, and molecules might go wrong on a much grander scale in the future. Despite what its proponents may claim for it, this is not an invincible or fail-safe technology.[3]

The Place of Genetic Engineering Biotechnology in Animal and Organic Agriculture

Recent advances in genetic engineering biotechnology are being developed for commercial application in animal agriculture. There are three basic approaches to enhance animal health and productivity using this new technology.

First, gene-spliced bacteria have been engineered to manufacture new generation animal vaccines, and also pharmaceuticals, like interferon to boost immunity and growth hormone to boost growth rates and milk yield. These latter products are claimed by manufacturers to be analogs of natural compounds already present in the animal's body. But safety and efficacy of these products of biotech "pharming" await verification. They are analogous (as distinct from homologous) products, i.e., not entirely natural. Their use in farm animals to artificially enhance immunity, disease resistance, growth rate, muscle mass, milk yield, etc., should be questioned in terms of overall animal health and well-being, since they will be utilized primarily in intensive, confinement-style animal production systems. Long-term social and economic consequences on the structure and future of agriculture are also considerable.

Their use in organic animal agriculture (with the exception of new generation genetically engineered vaccines) should be prohibited on the grounds that they are non-natural, analog products, and, with good breeding and husbandry practices, are unnecessary and unwarranted.

Second, gene-spliced microorganisms are being developed and will soon be marketed for feeding to pigs and poultry and for injection into the rumens of cattle to help improve feed digestibility and for reducing nitrate levels in manure. This material includes non-natural feed ingredients, such as sawdust and newspaper pulp. From an organic perspective, this is wholly unacceptable, no matter what efficiencies and cost savings might be claimed. To so alter the internal physiology of farm animals by bacterial manipulation is the antithesis of organic animal agriculture.

Third, gene-spliced farm animals are being developed, but their commercial future is at least 5 to 10 years away. By inserting the genes of other species or extra genes of their own kind into their developing embryos, so-called transgenic farm animals (and also fish) have been created. Some of them are able to transmit these additional genes to their offspring. With the exception of poultry (whose genetic lineage has been permanently changed by gene-splicing a segment of the fowl leukemia virus to convey immunity), these transgenic farm animals have been created to be either more productive, rather than disease or stress resistant, or to produce pharmaceutical products in their milk.

From an organic and holistic animal agriculture perspective, the creation of transgenic farm animals is an unacceptable and unnecessary alternative to traditional selective breeding and other humane, ecologically sound, and healthful husbandry practices.

Other developments in biotechnology include embryo-transfer, cloning and DNA mapping, which have been criticized as leading to a potential loss of genetic diversity in the farm animal populations and also to the selection of varieties of livestock and poultry that are suited only for intensive production systems.

Agribusiness biotechnologists reason that since one of the criteria for organic farming is the use of natural products (e.g., using natural rather than synthetic chemical fertilizers), then transgenic crop varieties should be eligible for organic certification since the foreign genes they contain from other life forms are natural in origin. This same line of reasoning would accept such genetically engineered products as biopesticides and bovine and porcine growth hormone as natural and thus acceptable under organic farming and food standards. This reasoning is flawed, however, because such bioengineered products and processes either do not naturally exist in conventional crops and animals that are part of an organic farming systems or occur at much lower concentrations within the normal homeostatic range of the plant or animals natural physiology and metabolism.

All developments in biotechnology should, from an organic farming and sustainable agriculture perspective, be subjected to evaluation on the basis of the following bioethical criteria:

1. Necessity. Is the new technology, product, or service really necessary, safe and effective; and are there alternatives of lesser risk and cost?

2. Public demand and acceptance.

3. Environmental impact, short-term and long-term, and influence on wild plant and animal (including invertebrate) species and microorganisms.

4. Release of genetically engineered life forms. Can they be identified, traced, contained or recalled if needed?

5. Economic impact, social justice, equity (international and intergenerational). Who will benefit? Who might be harmed?

6. Animal welfare. Will the new product or service enhance farm animal health and overall well-being?

7. Social and cultural consequences. How will it impact on the structure of agriculture, nationally and internationally, and on more sustainable traditional and alternative agricultural practices at home and abroad?

8. Oversight and compliance. Can the new technology, product, or service be effectively regulated to maximize benefits and minimize risks, and at what cost to society?

Conventional agriculture has a finite future because it is so heavily dependent on finite fossil fuels. This nonrenewable petrochemical resource base means that in the long run, conventional agriculture is nonsustainable. Agrichemical companies are now seeking and investing heavily in biotechnology alternatives. They claim that genetic engineering is the wave of the future and the way to feed the hungry world. But it is capital intensive even if, compared to agrichemicals, it is relatively more sustainable, if not also possibly safer.

From an ecological perspective it is not sustainable, because agribiotechnology is being misapplied as a profitable band-aid remedy for fundamentally unsound farming practices, most notably monocrop agriculture. These practices, not the genetic structure of crops and farm animals, need to be changed, with organic farming being the ideal goal. But low-input sustainable farming practices provide less profit for agribusiness which has a vested interest in maintaining a basically dysfunctional and nonsustainable industrial agro-ecology and agro-economy. A major barrier to the adoption of alternative sustainable agricultural practices is posed by the inappropriate use of biotechnology, such as in the engineering of herbicide-resistant and pesticide-producing plants, the patenting of crops, and the marketing of products like synthetic bovine growth hormone (rBGH).

For consumers to be able to help support organic farmers and humane sustainable agriculture, all foods must be labeled as to *how* and *where* they were produced, and whether or not genetic-engineering biotechnology was used in any phase of production or processing. Consumers' right to know is being undermined by agribusiness that would like to have the FDA treat all genetically engineered fruits, vegetables, and food additives as natural, safe, and not needing any special label. But fish antifreeze genes in tomatoes would violate the dietary code of ethical vegetarians; and foods with ingredients that are allergens for some people, like potatoes engineered to carry peanut protein genes, could cause anaphylactic shock and death.

In the final analysis, good government works with and for the people, and consumers' right to know about the food they purchase should be respected and upheld. While our political power may be limited under technocratic government, as consumers we have considerable power so long as we still have some choices in the market place. It's not what comes out of our mouths so much as what we put into them that really counts. The politics and wisdom of becoming an informed, conscientious consumer for one's own good, and for the good of all, are covered in the epilogue to follow.

Epilogue

Strategies for Change

Corporate and Government Initiatives

THE Cornucopia Project of the Regenerative Agriculture Association includes *A Survival Guide for Food Companies,*[1] which offers the following proposals. Rather than being retrogressive, as critics might contend, these proposals are progressive and speak to long-term profitability for the private and public sectors of our capitalist society alike, since, as democracy teaches, their well-being and future prosperity are mutually dependent. To serve the greater good of society is enlightened corporate self-interest.

> What should food industry executives be thinking about in order to begin adjusting to the difficult realities of our future? For starters, we'd suggest the following:
>
> *1. Long-term planning will become increasingly important.* No one can predict the future with total accuracy, but tracing the likely course of existing trends . . . makes some aspects of the future rather obvious. Without a long-range perspective, food companies may be overwhelmed by rapid changes, just as Detroit was nearly swamped by the oil crisis.

2. Food transportation will have to be kept to a minimum. Food isn't improved by being shipped, and the cost of transportation is hurting both companies and consumers. The food system of our future is likely to be much more regionalized, and tailored according to local needs, sources and markets. Companies that begin now to seek alternatives to cross-country food transport, and to adapt to regional patterns of production and sales, are likely to enjoy considerable success.

3. Dependence on fresh foods will increase. Bringing fresh, whole foods to market requires less energy and less capital than handling processed foods. In addition, consumers concerned about health will be purchasing more of these basic foods.

4. Seasonal eating patterns will return. During the age of food affluence, people felt they could afford any kind of food at anytime of the year. That period is now ending. Seasonal foods are more logical from both a production and a cost standpoint.

5. Consumers will continue to grow in sophistication and understanding. Shoppers will demand additional services from grocery stores, such as nutrition education, hints for easy food preparation, and directions for preserving fruits and vegetables. Others will seek to influence the way food is grown and marketed.

Rather than seeing these consumers as the enemy, food companies could look for opportunities to work with them toward the common goal of a more rational, sustainable food system.

6. Food will begin to be viewed in new ways. Most people are convinced that what they eat affects how they feel. As the costs of medical care increase, these people are taking more responsibility for their own health. A growing number will see food as a medicine, both to treat and to prevent disease. This will contribute to increased demand for fresh, wholesome, unprocessed foods.

A recent USDA survey showed that over the past three

years, two-thirds of U.S. households have changed their diets for health reasons, mainly to cut down on fats and cholesterol. The first casualty was beef. Consumption has dropped 16 percent since 1977.

7. *A revitalization in the discipline of home economics will occur.* There has been little interest in this field recently, and few innovative ideas. But now some new thinking is starting to infuse home economics.

The basic thrust is that consumers should get more value from existing investments in their home and property, and should see conservation as a source of family income. Industry, too, is beginning to see the importance of these ideas, which opens up another area for possible cooperation.

8. *Home food systems will offer new opportunities for some food marketers.* A home food system switches part of the processing of food from industry to the home. This movement is gaining momentum, because it saves consumers money, and gives them more control over their food. Natural and whole foods fit particularly well into this approach.

9. *Long range uncertainty about the economy and inflation will increase interest in food storage in the home.* People are reviving the arts of canning and drying, and freezer sales continue to increase. Again, the factors motivating this change are the desire to save money, and the security and satisfaction of being personally involved with your food supply. This tendency could have a big influence on consumer choices of both food types and packaging.

Wendell Berry offers the following "transformational steps":[2]

1. Withdrawal of public confidence "from the league of specialists, officials and corporate executives who for at least a generation have had almost exclusive charge of the problem and who have enormously enriched and empowered themselves by making it worse."

2. A change in values from labor-saving to recognizing the social and economic value of human energy.

3. Government should work to assure the equitable distribution of property through taxation. Today's tax structure undemocratically favors large corporate farms and industry.

4. Low-interest loans for those wishing to buy small family-size farms.

5. Price controls on production quotas to adjust production to needs and to the carrying capacities of farms, which would temper extreme fluctuations of supply (these work against small producers) and offset depressed prices at harvest time for small producers who cannot afford storage. Farmer cooperatives can help in this regard.

6. Programs to promote local self-sufficiency with local farmer and retailer market cooperatives. Repopulation of rural areas (especially in the Sun Belt) can help establish a more economical distribution of fresh food.

7. Local urban composting of uncontaminated sewage, garbage, and waste paper to be returned to the land at a nominal cost to farmers.

8. Revision of sanitation laws, which at present work against small dairies and packing plants, and favor overcentralized capital-intensive regional monopolies of food processing and distribution.

9. Land-grant schools should be required, as the Hatch Act instructs, "to insure agriculture a position in research equal to that of industry"; in other words, to encourage the greatest possible genetic and technological diversity in agricultural practices in accordance with local need to offset the present counterproductive uniformity in both categories.

10. A realignment of the overspecialized interests of agricultural colleges with those of local farmers and breaking their almost exclusive monopoly by agribusiness.

11. Development not only of diverse agricultural sys-

tems, but of systems consonant with the social, material, spatial, moral, spiritual, and economic principles of egalitarianism within the psychological dimensions of affective and effective human scale.

12. Development of a viable concept of absolute good (as distinct from but not exclusive of material progress) by which we can measure the value of our endeavors. That absolute good, Berry proposes, "is health—not in the merely hygienic sense of personal health, but the wholeness, finally the holiness, of Creation, of which our personal health is only a share."

13. Safety in diversity: Berry concludes that a highly centralized and industrialized food-supply system has no margins for error (as witness the PBB catastrophe in Michigan in 1975). A highly diversified, decentralized small-farm agriculture coupled with local marketing has more margins of safety. The myths that bigness is best and that competition will guarantee that only the most efficient agricultural systems will succeed need to be dispelled. Berry includes in his seminal treatise this poignant vision:

> Modern urban-industrial society is based on a series of radical disconnections between body and soul, husband and wife, marriage and community, community and the earth. At each of these points of disconnection the collaboration of corporation, government, and experts sets up a profit-making enterprise that results in the further dismemberment and impoverishment of the Creation.
>
> Together, these disconnections add up to a condition of critical ill health, which we suffer in common—not just with each other, but with all other creatures. Our economy is based upon this disease. Its aim is to separate us as far as possible from the sources of life (material, social, and spiritual), to put these sources under the control of corporations and specialized

professionals, and to sell them to us at the highest profit. It fragments the Creation and sets the fragments into conflict with one another. For the relief of the suffering that comes of this fragmentation and conflict, our economy proposes, not health, but vast "cures" that further centralize power and increase profits: wars, wars on crime, wars on poverty, national schemes of medical aid, insurance, immunization, further industrial and economic "growth," etc.; and these, of course, are followed by more regulatory laws and agencies to see that our health is protected, our freedom preserved, and our money well spent. Although there may be some "good intention" in this, there is little honesty and no hope.

Only by restoring the broken connections can we be healed. Connection is health.

Consumer Initiatives: Eating with Conscience[3]

A solution to Agricide is to eat with conscience. Every time we sit down to eat or purchase a food product, we are stating our preference and casting our vote for a specific type of agriculture, environment, and bioethics. We either make an informed choice for a humane, healthy, and sustainable food chain, or else we endorse "business as usual"—factory farming, chemical-intensive agriculture, and inferior food. The underlying philosophy to eating with conscience is that *every person has a right to safe, healthful food, as well as a clean environment, and that animals and the entire web of life and biodiversity deserve our respect and compassion.*

The goal of eating with conscience is simple, yet fundamental: seeking to promote, on a society-wide and global scale, food choices that are good for animals, good for the planet, and good for public health, and promoting those family farmers and ranchers who produce healthful food with compassion and conscience.

Every person has the right to decide what to eat, and they deserve to be fully informed about the environmental, health, and animal-welfare consequences and the hidden costs of food choices. The best strategy to achieve our goals is to launch a nationwide—indeed a worldwide—educational campaign. We welcome your participation in launching an Eating with Conscience campaign in your local community.

The typical American daily diet averages 37 percent of calories from fat, about 15 percent from protein, and 45 percent from carbohydrates.

The average American diet is too high in fat, protein, and cholesterol and too low in carbohydrates and fiber. The Dietary Guidelines for Americans, issued by the U.S. Department of Health and Human Services (HHS), state that we are consuming at least 7 percent more calories from fat than we should. Many other researchers and doctors advise trimming fat intake to 15–25 percent of caloric intake.

We are also consuming about one and a half times more protein than is recommended, and Americans take in a mere 45 percent, rather than the minimum recommended 50–60 percent, of calories from complex carbohydrates (starches found in grains, vegetables, and fruit).

Dietary factors are known to contribute substantially to the development of the following diseases, which are responsible for more than two-thirds of all deaths in the United States: coronary heart disease, stroke, atherosclerosis, diabetes, and certain cancers. A diet lower in fat, cholesterol, and protein and higher in fiber can help prevent these diseases.

Most American diets contain too much protein. Some of the excess is converted into fat, and too much protein may also contribute to disease. Epidemiological evidence has shown a correlation between increased protein consumption (particularly animal protein) and the incidence of such diseases as osteoporosis, cancer, heart disease, and diabetes.

Fat contains over twice the calories of an equal amount of carbohydrates or protein. Americans' overconsumption of fat in large part explains the dangerous obesity epidemic that afflicts a third of the population of our country. Obesity increases one's risk of high blood pressure, coronary heart disease, diabetes, and possibly some cancers as well as other chronic diseases.

Excess fat and cholesterol accumulate in human arteries and cause heart disease, which is the leading killer in the United States, causing one-third of all deaths. Our bodies produce all the cholesterol we need; any ingested cholesterol is in excess of what we need. Cholesterol is only found in animal products. Saturated fats—which are predominantly found in animal products—are the major dietary contributor to total blood cholesterol levels.

Fat is also implicated as a contributor to many of the cancers that afflict our population. Many pesticides, herbicides, and other chemicals are fat soluble and accumulate in animal fat. These are passed on to consumers in concentrated amounts, posing yet another health hazard.

Many respected health authorities, including the American Heart Association, the American Cancer Society, the National Cancer Institute, the National Institutes of Health, the National Academy of Sciences, the U.S. Department of Health and Human Services (HHS), the U.S. Department of Agriculture (USDA), and the Mutual of Omaha Insurance Companies recommend eating less fat and more fruits and vegetables.

Studies have shown that lowering the cholesterol level in diets can reduce heart disease and even reverse heart and artery changes that can lead to a heart attack. Mutual of Omaha, the nation's largest provider of health insurance for individuals, will reimburse heart patients who participate in a program to reverse heart disease that combines a very low-fat vegetarian diet, moderate exercise, stress management, and support groups.

Demographic studies have shown a strong association between fat intake and the incidence of breast, prostate, and colon cancer—the leading fatal cancers (along with lung cancer). According to the American Institute for Cancer Research, approximately 40 percent of all cancers in men and 60 percent of all cancers in women are associated with diet. By eating a diet lower in fat and higher in fruits and vegetables, you can substantially cut your risk from these killer cancers.

These are the key points of the dietary guidelines for Americans developed by the HHS:

/ Eat a variety of foods.
/ Maintain a healthy weight.
/ Choose a diet low in fat, saturated fat, and cholesterol.
/ Choose a diet with plenty of vegetables, fruits, and grain products.

The USDA and HHS have created the Food Guide Pyramid, in which they recommend bread, cereal, rice, and pasta as the basic components of one's diet. They also recommend that five or more servings of fruits and vegetables be consumed daily and that fats, oils, and sweets be used sparingly.

In addition to the previously mentioned chronic diseases, cases of acute foodborne illnesses are estimated as ranging from 6.5 to 81 million annually, with 9,000 fatalities. In most of these cases for which a source can be determined, animal products contaminated with such common bacteria as Salmonella, E. coli, Listeria, and Campylobacter caused the illnesses. The Food and Drug Administration (FDA) estimates that these acute illnesses cost the U.S. between $6 and $23 billion annually in medical expenses and lost productivity.

In the past four years, the Government Accounting Office (GAO) of the U.S. Congress has twice reported that the FDA and USDA do not adequately test meat and milk for antibiotic and other drug residues. A 1992 National Academy of Sciences report also found seafood inspection to be seriously inadequate, allowing dangerous pathogens and chemicals to enter the food supply.

A new drug, recently approved for use in dairy cows, is likely to result in the use of even more antibiotics. Recombinant (synthetic) Bovine Growth Hormone (rBGH, also known as rBST) is a genetically engineered hormone that has no therapeutic value and serves only to increase production of milk—a commodity that has long been in surplus. Already under stress from high-production demands, half of all dairy cows suffered from mastitis (infection of the udder) prior to the drug's approval. Because mastitis is one of the many listed side effects of rBGH, cows treated with rBGH may require even more antibiotics, the residue of which may end up in the milk and meat supply.

The vast majority of the animals used for food are kept in intensive confinement systems (factory farms). Animals living in these overcrowded and stressful conditions are susceptible to compromised immune systems and disease. To promote growth and prevent disease they are routinely fed antibiotics, which can lead to the development of antibiotic-resistant strains of bacteria that defy treatment in humans. Add to this risk the possibility that residues of these drugs may also end up in our food. Healthier living conditions for animals is the logical first line of defense against bacterial contamination and other food safety problems.

Some in the meat industry and the USDA have endorsed exposing meat to radioactive elements (irradiation) as the solution to the bacterial-contamination, food-safety crisis. However, this technology poses potential dangers. Does irradiation reduce the nutritional value of foods? Do irradiated foods contain dangerous byproducts of irradiation such as free radicals, which can damage human cells?

Irradiation provides an unacceptable excuse to continue employing irresponsible practices such as forcing animals to live in overcrowded and disease-promoting factory farms, washing poultry carcasses in contaminated baths, and running processing lines too fast for the USDA to adequately inspect the carcasses. Fast line speeds also lead to health and safety problems for plant workers—including repetitive motion disorders from having to perform the same task every two seconds. Irradiation is merely a stop-gap approach to a critically flawed meat production system.

There are no simple solutions. What is needed is the genuine commitment of industry, government, and consumers to food safety. This commitment must include better living conditions for farm animals, more humane handling of animals during transport and slaughter, cleaner and safer processing in slaughter plants, and safer food storage and preparation wherever food is served.

There is a better way to farm. Farmers practicing humane sustainable agriculture (HSA) fully satisfy farm animals' basic physical and behavioral requirements for health and well-being. Healthier animals require fewer drugs, in turn lowering consumer-health risks. Animal products from humane sustainable farms can be part of a more healthful diet. HSA products, animal and vegetable, come from healthy systems and soils farmed in ways that can be sustained over time.

Economic and Environmental Concerns

Agriculture uses an enormous amount of our natural resources, in large part to produce animal products. Almost half of the land in the United States is used for growing feed crops or for pasture and range. According to the USDA, nearly 40 percent of the world's grain products and 70 percent of U.S.-produced grain are fed to farm animals.

Converting feed into animal products is notoriously inefficient. For example, it takes almost seven pounds of grain and soy to produce one pound of pork. The amount of grain fed to farm animals could feed five times as many people if they were to consume it directly rather than in the form of animal products.

Animal agriculture also consumes a staggering amount of water. For instance, to produce just one pound of grain-fed meat requires from 300 to 500 gallons of water, and U.S. poultry operations use 96.5 billion gallons of water annually. To supply a person with meat, milk, and eggs each day requires one hundred gallons of water per day, which is estimated to be as much as is used for all other household purposes combined.

In addition, agriculture uses up vast amounts of energy, with meat leading in the amount of energy used per pound of product served. For example, it takes fifteen times more energy to get pork to the consumer's plate than fresh fruit and vegetables. Cornell University Professor David Pimentel has estimated that if the diet of the world's population was obtained using methods as energy intensive as those of the United States, all known oil reserves would be consumed in merely twelve years.

Not only does contemporary animal agriculture use up enormous amounts of precious resources, it also pollutes and destroys them. The Environmental Protection Agency (EPA) has classified agriculture as the leading cause of water pollution, which results in serious consequences for the wildlife and ecology of lakes and rivers. One-third of nonpoint pollution (pollution originating from an unspecified source) consists of animal waste. About 2 billion tons of animal manure are generated each year in the United States alone. Agricultural chemicals are primary sources of water pollution and have been found in the groundwater of most states. Among other problems, this can adversely affect human and animal health by contaminating drinking water.

Additionally, an estimated 100 million acres of U.S. cropland have already been severely degraded and abandoned. Farm animals themselves are not inherently environmentally destructive, but rather the prevailing methods used to obtain animal products are. Animals on traditional family farms have usually been ecologically well integrated. For the last 50 years there has been a trend toward replacing traditional family farms with corporate-owned or contracted factory-like animal-production systems. The objective of factory "farms" is to keep huge numbers of animals together, each confined in as small a space as possible, in the name of efficiency. The vast majority of farm animals now spend their entire lives crowded, in some cases by the tens of thousands, inside buildings that are designed for maximum productivity rather than for the animals' health and well-being. Millions of farm animals are kept in feedlots—large outdoor corrals, often with no shade or shelter.

Concentrating so many animals together in this way breaks the ecological cycle and causes suffering and disease. Furthermore, animal wastes are transformed from valuable soil nutrients into hazardous waste because there are simply too many animals for the surrounding land to accommodate.

Responsible grazing management, practiced by some ranchers, can actually improve land. While more extensive outdoor systems are advocated, inappropriate practices can be gravely problematic. Approximately half of the western United States is publicly owned and, of this, roughly 75 percent of the land is leased for grazing. This, in addition to other publicly and privately grazed lands, produces a total of 70 percent of the West being open to grazing. This region is a fragile environment where regeneration occurs slowly. In contrast to the region's native wildlife, domesticated cattle and sheep have destructive grazing patterns. After more than a century of unchecked grazing, 35 percent of U.S. grazing land has been severely desertified. The West cannot sustain the millions of domestic animals grazing it.

In addition, the USDA's Animal Damage Control program used tens of millions of tax dollars to poison, trap, and shoot millions of coyotes, foxes, and other wild animals last year. Many were killed purportedly to protect cattle, sheep, and crops.

Cattle from Central American rain forests are primarily used as a source of cheap beef for fast-food restaurants in the United States. Converting forests to grazing land destroys wildlife habitats, endangering an indeterminable number of animal and plant species. Soils are very thin in most tropical forests, and once cleared they can support cultivation for only a few years before rains leach out nutrients and erode the vital topsoil, laying the land to waste. Almost two-thirds of Central America's rain forests have already been cleared or severely degraded, in large part as a result of cattle ranching, which is often subsidized by the governments of those countries.

The great demand for animal feedstuffs encourages people in Third World countries to abandon sustainable farming practices in order to grow cash crops for export. Profits are short- lived, however, as the marginal lands quickly erode, contributing to poverty, malnutrition, and the loss of wild lands.

There are farmers and ranchers who work to make a living in cooperation with their land and animals, rather than treating the soil like "dirt" and animals as if they were commodities. These farmers practice humane sustainable agriculture (HSA), which produces adequate amounts of safe, wholesome food in a manner that is **ecologically sound, economically viable, socially equitable, and humane.**

Farmers practicing HSA emulate natural cycles and recycle animal wastes and other crop nutrients on their farms. They rely more on their own on-farm resources than on externally derived ones. Animals are given sufficient room for normal behavior and are kept at appropriate population densities so that their wastes are not in excess of that which can be accommodated by the surrounding land. HSA producers employ rotational grazing and other management systems that restore fertility to worn-out pastures without polluting waterways.

Returning grazing animals to grass-feeding systems, and away from the prevailing grain-based systems, would reduce the amount of energy required by about 60 percent, and the amount of land required by about 8 percent. An ample amount of protein would still be produced, while 300 million tons of grain would be made available.

If not overstocked, land that is unsuitable for crop production can support cattle, sheep, and goats, all of which convert the otherwise unusable cellulose in grass and shrubs into protein. Holistic range management and rotational grazing methods, whereby appropriate breeds and numbers of animals graze seasonally, can actually improve rangeland, while guard dogs, herders, and other nonlethal methods can protect domestic animals from predators. Geese, chickens, and hogs can also play a vital role in weed and pest control in ecological farming systems.

The adoption of sustainable agricultural practices can reduce the use of fertilizers, pesticides, irrigation, and energy, making agriculture environmentally sound while at the same time making our food supply a safer one, improving conditions for animals, and increasing profits for farmers.

Farm Animal Welfare

While it may surprise many people, the animals most commonly used for food (chickens, cows, and pigs) are complex, sensitive, highly social animals. They retain most of the needs and behavioral drives of their ancestors and wild relatives. Along with virtually every other animal, including humans, they share the need for personal space—a minimum distance that they try to keep between themselves and others. Like dogs, cats, and all animals, farm animals can experience stress, frustration, fear, pain, and pleasure. It logically follows that many of the concerns we have for the well-being of other animals also hold true for farm animals.

Today, however, the vast majority of farm animals are kept in "factory" farms—systems of extreme confinement in which the greatest possible number of animals are raised in the smallest possible space and natural behavior is impossible.

If we keep and use animals for food, we should care enough to ensure that they are treated humanely. Choosing alternatives to animal products that come from inhumane systems will help accomplish this.

For the last few decades, as family farms have been forced out of business by large agribusiness corporations, the trend has been toward keeping animals in factory-farm systems. Since there are no federal laws regulating the treatment of animals on farms, managers are free to use whatever methods of production are most profitable, regardless of their impact on the animals' well-being.

Factory farming occurs behind walls and fences, usually far from public view. Removed from contact with farm animals, people retain the storybook images they encountered as children or accept as truth advertising's images of contented animals roaming freely in barnyards and pastures. Most people have no idea of how farm animals are really treated today.

Industry's claim that a productive animal is a healthy one is deceptive. Selective breeding and drugs (including antibiotics) and other chemicals are used to force animals to produce, despite adverse conditions, at the expense of their health. New diseases have been brought about by exhaustive production demands and by inappropriate diets, breeding practices, and housing.

Dust and ammonia and other gases accumulate in factory-farm systems, causing chronic respiratory disease and death. Agricultural animal disease costs in the United States total $17 billion per year. Such huge losses are considered acceptable because profits depend on overall output and the optimal use of space and equipment, not on the well-being of individual animals.

Some factory-farm systems mentioned, including battery cages for hens and crates for sows and calves, are banned or are being phased out in other countries. As more and more people learn the facts about how these animals are being treated, they add their voices to the growing demand for change.

Supporting Humane Sustainable Agriculture

There are farmers and ranchers who care about the well-being of their animals, the environment, and the quality of the food they offer. They practice humane sustainable agriculture (HSA), which The HSUS has long supported. HSA consists of food-production systems and practices that meet the physical and behavioral needs of animals while treating natural resources in ways that ensure continued productivity for the generations to come.

America's unnecessarily destructive and excessive production of meat and other animal products, and the subsequent overconsumption of them, causes many of our health problems as well as those of our environment. A change to more humane and sustainable systems can lead to a revolution in our nation's food system that will benefit all. More dollars flowing to humane sustainable farms will encourage others to farm this way, while less money spent on monoculture crops and factory-farmed products will make them less economically viable.

To help you eat with conscience and choose a more humane diet, The HSUS encourages you to consider the three R's when you shop or dine out:

Reduce your consumption of animal products, for your own health as well as that of the environment.
Replace those animal products in your diet with grains, fruits, and vegetables.

Refine your diet by purchasing products obtained from more humane and sustainable systems. Shop at farmers' markets, health food stores, and food co-ops, all of which are likely to carry organic, humane and sustainable foods. Look for words like "free-range," "free-roaming," "free-running," or "uncaged" on meat and egg labels. *The Humane Consumer and Producer Guide*, available from The HSUS, will help you locate sources of these products.

Demand that any dairy products you buy be obtained from cows that have not been injected with recombinant (synthetic) Bovine Growth Hormone (rBGH). Get involved in the efforts of public-interest groups to require that food be labeled in terms of how and where it was produced and processed.

We encourage you to join with us in the transition to an agriculture that is better for animals, better for the environment, and Good for You!

For more information contact:

Howard Lyman
Director, Eating With Conscience Campaign
The Humane Society of the United States
2100 L Street, NW
Washington, D.C. 20037
301-258-3054
301-258-3081 Fax
hlyman·aol.com E-Mail

The Vegetarian Imperative

It may seem offensive to vegetarians that I endorse the concept of treating farm animals humanely because it would seem that I support the livestock and poultry industries. Actually, I don't. I am—for reasons espoused by such people as the Buddha, Pythagoras, Plato, Benjamin Franklin, John Wesley, Leo Tolstoi, Charles Darwin, George Bernard Shaw,

Mohandas Gandhi, Albert Einstein, and, not unnaturally, Upton Sinclair—a vegetarian.

Treating farm animals humanely and not like machines entails a fundamental change in attitude. This same attitude needs to be applied to the way in which the land—nature and her resources—is treated. We suffer if we do not live by the principles of the higher self: exercising humility and compassion in our relationships with all living things and acquiring empathetic (rather than instrumental) understanding of how nature works and how our world view and actions can harm or benefit ourselves and other living things. We cannot harm them without ultimately harming ourselves, be it from treating farm animals inhumanely or from killing life-transforming bacteria in the soil.

Conversely, we cannot but benefit ourselves when we benefit other living things. This happens when we begin to live in reverence for all life, shifting from an adversarial mentality of conquest of nature and war against pests and disease to one that sees the wisdom of planetary and environmental restoration and of living in harmony with nature. All of life is sacred, since all life is part of the One Life. And if we continue to treat each other, other animals, and the environment without that kind of reverence, we will suffer the consequences—economically, ecologically, physically, and spiritually.

Professor Arthur Westing estimates that

> in a mere four decades or so, the human *plus* livestock biomass is likely to represent fully 40 percent of the total animal biomass—all of this being at the inevitable expense of more than an equivalent biomass of wildlife. Even if livestock could be raised more efficiently than wildlife, such gains would be more than negated by habitat losses resulting from urbanization, etc., associated with human population-increases.[4]

Vegetarianism is a personal choice. It implies a significant contribution toward restoring U.S. agriculture and enhancing our relationship with animals. If we do not apply the medical maxim "do no harm" to our treatment of animals and the land, and extend the Golden Rule to embrace all living things, then we shall suffer with the rest of creation which, as St. Paul said, "groaneth and travaileth until now."

The Christian view on the practical and ethical imperative of changing American dietary habits so as to consume less meat is eloquently expressed in *Earthkeeping: Christian Stewardship of Natural Resources:*

> A concern, both for the world's poor and for the maintenance of the healthy diversity of the ecosphere, suggests that large-scale changes are needed in the diet to which most North Americans have been accustomed. Are we willing to reduce our own personal impact on the food production system? For reasons of economic and political survival, we, or our children, *might* have to. But for reasons of compassion, love, and stewardly responsibility, not to mention for reasons of personal health, we *might* want to reduce our consumption. In either case some sort of change in diet seems inevitable for North Americans in the near future. As Christians, we must determine whether that change will be forced upon us, or whether we will be leaders in effecting it.[5]

Now is the time to work toward ending all forms of domination over nature and the animal kingdom—unless we wish to continue to demean our humanity by acting as unnaturally destructive and wasteful predators rather than as planetary stewards. There is no reason, other than greed to treat farm animals inhumanely. To think about farm-animal welfare and to adopt humane husbandry alternatives is a step toward animal liberation and ultimately toward our own liberation. We must strive to unfetter the animal kingdom from our cruelly destructive dominion. The fewer farm animals that are bred and raised, and the more wildlife and wild plants are reintroduced as we regenerate this tired and poisoned planet, the healthier our children shall be.

Masanobu Fukunoka, a Japanese farmer-philosopher, sees that "the ultimate goal of farming is not the growing of crops but the cultivation and perfection of human beings."[6] And Wes Jackson observes, "It is the human agricultural system which must grow more toward the ways of nature, rather than the other way round. . . . We have yet to develop an agriculture as sustainable as the nature we destroy."[7]

I see vegetarianism and organic, humane sustainable agriculture as significant as steps toward a Golden Age to come, wherein high-tech innovations such as *appropriate* genetic engineering (by which we will be able to manufacture many essential nutrients with the aid of engineered microorganisms) will help humanity survive and prosper, but not at nature's expense or in violation of the rights of all living things to a whole and healthy environment and to equal and fair consideration. Regardless of whether or not animals have a right not to be eaten, far too many are being produced and consumed. And this is at the expense of their welfare (since it is less economical under today's "factory" conditions to raise them humanely); at the expense of agricultural diversity and natural resources (especially topsoil and water); at the expense of wildlife (whom they and the land used to raise their feed and forage displace); and at the expense of consumer health. Vegetarianism is therefore an enlightened decision, if not a survival imperative in the long term, and producing and consuming less farm-animal produce, especially meat, are essential steps toward the overall restoration of our culture, agriculture, and environment.

Notes

Chapter 1: The Farming Business

1. *Animal Factories* (New York: Crown, 1981).
2. T. Cunha, in A. M. Altschul and H. L. Wilcke, eds., *New Protein Foods*, vol. 3, pt. A, *Animal Protein Supplies* (New York: Academic Press, 1978).
3. USDA, *Agricultural Statistics* (Washington, D.C.: GPO, 1978).
4. Testimony before the House Government Operations Subcommittee, July 24, 1985.
5. Office of Technology Assessment, *Drugs in Livestock Feed*, vol. 1, *Technical Report* (Washington, D.C.: GPO, 1979).
6. Comptroller General of the United States, *Problems in Preventing the Marketing of Raw Meat and Poultry Containing Potentially Harmful Residues* (Washington, D.C.: General Accounting Office, 1979).
7. Council for Agricultural Science and Technology, *Antibiotics in Livestock Feed*, Report 88 (Ames, Iowa: CAST, 1981), 9.
8. Ibid., 10.
9. "Current Status of Feed Additives in Livestock Production." Paper presented at Farm Science Day, Purdue University, January 14, 1971. Published in *Proceedings of the Meat Industry Research Conference, University of Chicago, March 23–24, 1972*.
10. Ibid., 1.
11. C. Burbee, "Antibiotic Feed Additives: The Prospect of Doing Without," *Farmline* 1:9 (December 1980).
12. "Medical News," *Journal of the American Medical Association* 242:14 (October 5, 1979): 1464.
13. *Federal Register* 43, no. 14:3034; cited in USDA, *Economic Effects of a Prohibition on the Use of Selected Animal Drugs*, Agricultural Economic Report 414 (November 1978), 3.
14. U.S. International Trade Commission, *Synthetic Organic Chemicals: United States Production Sales*, Publication 804 (Washington, D.C.: GPO, 1975); cited in USDA, *Economic Effects of a Prohibition*, 4.

15. USDA, *Economic Effects of a Prohibition,* 3–4.
16. Krider, "Current Status of Feed Additives."
17. USDA, *Economic Effects of a Prohibition.*
18. Ibid.
19. D. Pimental, "Grass-fed Livestock Potential: Energy and Land Constraints," *Science* 207 (1980): 843–48.
20. Louis H. Bean in Altschul and Wilcke, *New Protein Foods.*
21. G. A. Borgstrom, editorial in "Proper Diet Saves Lives, Land, Oil . . . ," a symposium held by the American Association for the Advancement of Science at Toronto, Canada [hereafter: "Proper Diet"], reported in *Science News* 119 (1981), 39–40.
22. D. Pimental, in "Proper Diet."
23. *Empty Breadbasket? The Coming Challenge to America's Food Supply and What We Can Do About It: A Study of the U.S. Food System,* the Cornucopia Project of the Regenerative Agriculture Association (Emmaus, Pa.: Rodale Press, 1981).
24. USDA, *A Time to Choose: Summary Report on the Structure of Agriculture* (Washington, D.C.: GPO, 1981).
25. *Empty Breadbasket?*
26. Ibid.
27. Ibid.
28. Ibid.

Chapter 2: The Economic Dimension

1. *New York Times,* August 25, 1985.
2. *National Hog Farmer,* March 15, 1984.
3. C. Davenport et al., *The Effects of Tax Policy on American Agriculture,* USDA, Agriculture Economic Report 480 (Washington, D.C.: GPO, 1982).
4. Ward Sinclair, "By Hawking Our Grain We Gamble Our Future," *Washington Post,* September 25, 1984.
5. *Washington Post,* May 20, 1982.
6. Mason and Singer, *Animal Factories.*
7. Ward Sinclair, "Pesticide Industry Getting Way on New Law," *Washington Post,* March 24, 1982.
8. American Farm Bureau Federation 1981 policy booklet.
9. *Scientific Aspects of the Welfare of Food Animals,* Report 91 (Ames, Iowa: CAST, 1981).
10. M. W. Fox, *Farm Animals: Husbandry, Behavior and Veterinary Care. Viewpoints of a Critic* (Baltimore: University Park Press, 1983).
11. J. K. Wise, *Journal of the American Veterinary Medical Association* 171 (1978): 1064.
12. *Journal of the American Veterinary Association* 184 (1984): 1289.
13. J. F. Klienbenstein, C. L. Kirtley, and M. L. Killingsworth, in

Swine Research, University of Missouri College of Agriculture, Special Report 273 (Columbia, Mo.: University of Missouri, 1979), 161–64.

14. W. S. Koski, "Marketing Food Animal Medicine," *D.V.M. Magazine* 13:2 (1982), 57–59.

15. "Perspectives of Integrated Pest Management," *Crop Protection* 1 (1980): 5–26.

16. Klienbenstein et al.

17. *National Hog Farmer,* March 15, 1984.

18. In *Southern Exposure: Our Food, Our Common Ground,* a bimonthly publication of the Institute for Southern Studies (P.O. Box 531, Durham, N.C. 27702), 9, no. 6 (November/December 1983).

19. See Ward Sinclair and Martha M. Hamilton, "Tax Laws Effect on Agriculture under Scrutiny," *Washington Post,* May 28, 1984.

20. M. Gabel, *Ho-Ping: Food for Everyone* (New York: Anchor, 1979).

21. *The Poverty of Power* (New York: Knopf, 1976).

22. K. J. Donham, M. Rubino, T. D. Thedell, and J. Kammermeyer, "Potential Health Hazard to Agricultural Workers in Swine Confinement Buildings," *Journal of Occupational Medicine* 19 (1977): 383–87.

23. USDA, *Fact Book of U.S. Agriculture* (Washington, D.C.: GPO, 1981), 34.

24. C. F. Nuckton et al., "Farm Size and Community Welfare: An Interdisciplinary Approach," *Rural Sociology* 47 (1982): 32–46.

25. *The Unsettling of America: Culture and Agriculture* (San Francisco: Sierra Club Books, 1977).

26. *Empty Breadbasket?*

27. *New York Times,* August 18, 1985.

28. Sinclair and Hamilton, "Tax Laws Effect."

29. *Empty Breadbasket?*

30. Laura King and E. Phillip LeVeen, *Turning Off the Tap on Federal Water Subsidies: The Central Valley Project: The $3.5 Billion Giveaway.*

31. Associated Press, August 21, 1985.

Chapter 3: The Health of the Earth and Its People

1. Sandra S. Batie, "How Farmers Affect Natural Resources," in USDA, *Using Our Natural Resources: 1983 Yearbook of Agriculture* (Washington, D.C.: GPO, 1983), 405.

2. *Empty Breadbasket?*

3. Ibid.

4. In Altschul and Wilcke.

5. *Empty Breadbasket?*

6. Dr. Norman Myers, *BBC Wildlife* 2 (February 1984).

7. Personal communication.

8. Paper delivered to the 49th North American Wildlife and Natural Resources Conference, 1984.

9. American Farm Bureau Federation, *Handbook of Policy Positions* (Washington, D.C.: AFBF, n.d.).

10. April 1985, 14.

11. *Science* 228 (1985): 1073–74.

12. "The Holy Cow—Provider or Parasite? A Problem for Humanists," *Southern Humanities Review* 13 (1978): 251–78.

13. F. H. King, *Farmers of Forty Centuries: Permanent Agriculture in China, Korea, and Japan* (Emmaus, Pa.: Rodale Press, 1973).

14. American Farm Bureau Federation, *Handbook*.

15. U.S. Bureau of Reclamation, San Luis Unit, Central Valley Project, California, Information Bulletin 2, February 1984.

16. Larry Ephron, in *Acres USA*, September 1983.

17. George A. Borgstrom in *Science News* 119 (1981): 39–40.

18. *Proteins: Their Chemistry and Politics* (New York: Basic Books, 1968).

19. *The Food and People Dilemma* (Boston: Duxbury Press, 1970).

20. *Acres USA*, July 1983.

21. *Man's Plague? Insects and Agriculture* (Princeton, N.J.: Darwin Press, 1976).

22. USDA, *Economics of Scale in Farming (Washington, D.C.: GPO, 1967).*

23. *Empty Breadbasket?*

24. *Selected Papers by John Hamaker,* annotated by Donald A. Weaver (Hamaker-Weaver Publications, P.O. Box 1961, Burlingame, Calif., 1984). See also the review by Larry Ephron, "How Soon Will We Have to Stop Eating?" *Acres USA*, May 1984, 10–13.

25. Office of Technology Assessment, *Drugs in Livestock Feed*.

26. Comptroller General of the U.S., *Problems in Preventing the Marketing of Raw Meat and Poultry*.

27. See D. L. Davis and B. H. Magee, "Cancer and Industrial Chemical Pollution," *Science* 206 (1979): 1356.

28. *New Scientist*, August 5, 1982, 351.

29. *Washington Post*, June 9, 1984.

30. Cass Peterson, in the *Washington Post*, June 30, 1984.

31. *Science News* 114 (1978): 8.

32. M. Dover and B. Croft, *Getting Tough: Public Policy and the Management of Pesticide Resistance* (Washington, D.C.: World Resources Institute, 1984).

33. D. Pimental et al., *Benefits and Costs of the Pesticide Use in U.S. Food Production* (Ithaca, N.Y.: Cornell University Press, 1980).

34. Harry Walters, "Nitrite and Cancer—A Broader View," *The Ecologist* 14 (1984): 32–37.

35. R. P. Sharma and J. C. Street, *Journal of the American Veterinary Medical Association* 177 (1980): 149.

36. J. B. Ludvigsen et al., in *Livestock Production Science* (Amsterdam: Elsevier) 9 (1982): 65–87.

37. *Feedstuffs*, July 2, 1979.

38. *Mother Jones*, January 1983.

39. Carmen A. Saenz Rodriguez, "Environmental Hormone Contamination in Puerto Rico," *New England Journal of Medicine* 310 (1984): 1741–42.

40. *New Scientist*, April 27, 1978, 211.

41. Reported in *Feedstuffs*, February 27, 1978.

42. Seth S. King, in the *New York Times*, April 22, 1979.

43. National Academy of Sciences, *Meat and Poultry Inspection: The Scientific Basis of the Nation's Program* (Washington, D. C.: National Academy Press, 1985).

44. *Veterinary Medicine* 79 (1984): 254–59.

45. *Feedstuffs*, March 19, 1984.

46. *Moneysworth*, March 1978.

47. *The Effects on Human Health of Subtherapeutic Use of Antimicrobials in Animal Feeds* (Washington, D.C.: National Academy of Sciences, 1982).

48. Published by and available free from Feed Additives Department, Animal Health Division, American Hechst Corporation, Somerville, N.J. 08876.

49. T. F. O'Brien et al., "Molecular Epidemiology of Antibiotic Resistance in Salmonella from Animals and Human Beings in the United States," *New England Journal of Medicine* 107 (1982): 1–6.

50. Reported in *Science* 213 (1981): 848.

51. T. F. O'Brien et al., in *Science* 230 (1985): 87–88.

52. W. G. Crook, *The Yeast Connection: A Medical Breakthrough* (New York: Professional Books, 1984).

53. *Veterinary Record* 102, no. 39 (1978).

54. *Veterinary Record* 104, no. 23 (1979).

55. *American Journal of Veterinary Research* 45 (1984): 1245–49.

56. S. D. Holmberg et al., in *Science* 229 (1984): 833–35.

57. S. D. Holmberg et al., in the *New England Journal of Medicine* 311 (1984): 617–622.

58. *Science* 229 (1984): 833–35.

59. "Playing Antibiotic Pool: Time to Tally the Score," *New England Journal of Medicine*, 1984. See also "News and Comment: Use of Antibiotics in Animal Feed Challenged," *Science* 226 (1984): 144–46.

60. J. F. Ferrer, S. J. Kenyon, and P. Gupta, "Milk of Dairy Cows Frequently Contains a Leukemogic Virus," *Science* 213 (1981): 1014–16.

61. K. J. Donham, J. W. Berg, and R. S. Sawin, "Epidemiologic Relationships of the Bovine Population and Human Leukemia in Iowa," *American Journal of Epidemiology* 112 (1980): 80–92.

62. See Dr. Virginia Livingston-Wheeler with Edmond G. Addeo, *The Conquest of Cancer* (New York: Franklin Watts, 1984).
63. Summarized in *Acres USA*, May 1984.
64. Ibid.
65. *New York Times*, September 9, 1985.

Chapter 4: The Nutritious Diet

1. "Proper Diet."
2. Gabel, *Ho-Ping*.
3. F. Chaboussou, *Les Plantes Malades des Pesticides: Bases Nouvelles d'une Prévention Contre Maladies et Parasites* (Paris: Debard, 1980).
4. Fox, *Farm Animals*.
5. R. J. Williams et al., "The 'Trophic' Value of Foods," *Proceedings of the National Academy of Sciences* 70 (1973): 710–13.
6. A. Tolan et al., in the *British Journal of Nutrition* 31 (1974): 185–87.
7. *Food Chemical News*, April 28, 1980, 20.
8. *American Heart Association Journal*, May 1984.
9. See H. Aldercreutz et al., in *Cancer Research* 41 (1981): 3771–73, and *Endocrinological Cancer, Ovarian Function and Disease* (Amsterdam: Elsevier, 1979).
10. "Proper Diet."
11. American Farm Bureau Federation, *Handbook*.
12. *The XO Factor* (New York: Park City Press, 1984).
13. "Don't Drink Your Milk," *East-West Journal*, July 1978, 35–40.

Chapter 5: The Matter of Conscience

1. *All That Dwell Therein* (Berkeley: University of California Press, 1982).
2. Ibid.
3. *Animal Liberation* (New York: Random House, 1975).
4. M. W. Fox, *Returning to Eden: Animal Rights and Human Responsibility* (New York: Viking, 1980).
5. *Parity Foundation News Magazine* 1 (1981): 7–13.
6. D. R. Griffin, *The Question of Animal Awareness* (New York: Rockefeller University Press, 1981).
7. J. R. Dawson, in *Confinement*, January 1977.
8. B. L. Moore and P. J. Chenoweth, *Grazing Animal Welfare* (Australian Veterinary Association, Queensland Branch, P.O. Box 34, Indooroopilly, Queensland, Australia 4068, 1985).
9. Fox, *Farm Animals*.
10. D. Carpenter et al., *Animals in Ethics* (London: Watkins, 1980).
11. George A. Borgstrom, in "Proper Diet."
12. *Federal Register* 48, no. 12 (January 18, 1983).

Chapter 6: Measuring the Results

1. See also J. P. Madden, *Economics of Size in Farming*, USDA Agricultural Economic Report 107 (Washington, D.C.: GPO, 1967); M. Kramer, *Three Farms: Making Milk, Meat and Money from the American Soil* (Boston: Little, Brown, 1980); and Gabel, *Ho-Ping*.
2. G. A. Benson and S. R. Sutter, *The New Farm*, January 1981, 4.
3. "The Economics of Small Farms," *Science* 219 (1983): 1037–41.
4. *Environmental Management in Animal Agriculture* (Urbana, Ill.: Animal Environment Services, 1981).
5. *New Scientist*, March 28, 1985.
6. V. R. Eidman and D. D. Greene, "An Economic Analysis of Three Confinement Hog Finishing Operations," University of Minnesota Agricultural Experiment Station Bulletin 535, 1980.
7. *Environmental Management*.
8. January–February 1979, 6–7.
9. Cited in *Michigan Farmer*, March 17, 1984.
10. "Perspectives of Integrated Pest Management."
11. Pimental et al., *Benefits and Costs*. See also D. Pimental and C. A. Edwards, "Pesticides and Ecosystems," *Bioscience* 32 (1982): 595–600.
12. *Empty Breadbasket?*
13. Pimental, in "Proper Diet."
14. *Empty Breadbasket?*
15. "Energy Costs of Intensive Livestock Production," paper no. 75-4042, American Society of Agricultural Engineers, June 1975.
16. Pimental, "Proper Diet."
17. Pimental, "Grass-fed Livestock Potential."
18. Schwabe, "The Holy Cow."
19. J. T. Reid and O. D. White, in Altschul and Wilcke.
20. "Proper Diet."
21. D. K. Britton and B. Hill, *Size and Efficiency of Farming* (Farnborough, England: Saxon House, 1975).
22. "Is Factory Farming Really Cheaper?"
23. *An Environmental Primer* (New York: Schocken, 1973).
24. C. S. Williams, quoted in *Feedstuffs*, April 23, 1984, 8.
25. Personal communication.
26. "Productivity Comparisons Between Thompson's Gazelle and Cattle and Their Relation to the Ecosystem in Kenya," Ph.D. thesis, Cornell University, 1975.

Chapter 7: Toward a Saner Future

1. D. Avery, "U.S. Farm Dilemma: The Global Bad News Is Wrong," *Science* 230 (1985): 408–12.
2. Associated Press, September 1985.

3. Jack Doyle, *The Altered Harvest* (New York: Viking, 1985).

4. Office of Technology Assessment, *Technology, Public Policy, and the Changing Structure of American Agriculture: A Special Report for the 1985 Farm Bill* (Washington, D.C.: GPO, 1985).

5. Planet Drum Foundation, P.O. Box 31251, San Francisco, Calif. 94131.

Chapter 8: Breaking the Cycle of Poverty and Famine: The Role of Humane Sustainable Agriculture

1. From The population, agriculture and environment nexus in *Sub-Saharan Africa.* Draft for discussion, May 29, 1990, p. 13. (Washington, DC: World Bank, Africa Region).

2. In *Our Country, The Planet.* (London: Lime Tree Press, 1992). See also Donella H. Meadows and Jorgen Randers, *Beyond the Limits.* (London: Earthscan, 1992).

3. G. M. Ward, T. M. Sutherland, and J. M. Sutherland, "Animals as an energy source in third world agriculture," *Science* 208 (1980): 570–75.

4. A. B. Carles, "The non-medical prevention of livestock disease in African rangeland ecosystems," *Preventive Veterinary Medicine* 12 (1992): 165–73.

5. T. J. Cunha, et al., *Animal Productivity. Supporting Papers: World Food and Nutrition Study,* vol. 1, (Washington, DC: National Academy of Sciences and National Resource Council, 1977).

6. See T. R. Preston and E. Murgueitio, (1992) *Strategy for Sustainable Livestock Production in the Tropics.* CIPAV/SAREC, PO Box 16140, S-103 23 Stockholm, Sweden (CIPAV-Carrera 35A, Oeste #3–66, AA20591 Cali, Columbia).

7. C. Reijntjes, B. Haverkort and A. Waters-Bayer, *Farming for the Future: An Introduction to Low-External Input and Sustainable Agriculture* (London: Macmillan Press Ltd., 1992).

8. For further discussion see Jeremy Cherfas, "Farming goes back to its roots," *New Scientist,* May 9, 1992.

9. Fred Pierce, "The hidden costs of technology transfer," *New Scientist,* May 9, 1992.

10. For further discussion on impact of agricultural biotechnology on the Third World see Vandana Shiva, *Monocultures of the Mind* (Penang Malaysia Third World Network, 1993).

11. Jonathan M. Harris, *World Agriculture and the Environment.* (New York: Garland, 1990).

12. L. and B. A. Stewart, *Soil Degradation,* (New York: Garland, 1990). See also D. Pimental, 1992. *World Soil Erosion and Conservation* (Cambridge, UK: Cambridge University Press, 1992).

13. *Facts and Figures: International Agricultural Research.* (New York: Rockefeller Foundation and IFPRI. 1990).

14. M. B. Green, H. M. Le Baron, and W. K. Moberg, *Managing Re-*

sistance to Agrochemicals: From Fundamental Research to Practical Strategies (Washington, DC: American Chemical Society 1990).

15. Winrock International Institute for Agricultural Development, Route 3, Box 376, Morrilton, Arkansas 72110-9537.

16. John S. Mbiti, *Introduction to African Religion* (Oxford, England: Heinemann Educational Books, Ltd., 1991).

Chapter 9: Genetic Engineering and Our Farming Future

1. Michael W. Fox, *Superpigs and Wondercorn: The Brave New World of Biotechnology and Where It All May Lead* (New York: Lyons and Burford, 1992).

2. See Cary Fowler et al., *The Laws of Life. Another Development and New Biotechnologies Development Dialogue* (Uppsala: Dag Hammerskjold Foundation, 1988), 1–2.

3. Jack Doyle, *The Altered Harvest*, (New York: Viking 1985).

Epilogue

1. The Cornucopia Project of the Regenerative Agriculture Association, *A Survival Guide for Food Companies* (Emmaus, Pa.: The Cornucopia Project, 1981).

2. *The Unsettling of America.*

3. Coauthored with Howard Lyman.

4. "A World in Balance," *Environmental Conservation* 8 (1981): 177–83.

5. Edited by Loren Wilkinson (Grand Rapids, Mich.: Eerdsman, 1980).

6. *The One Straw Revolution* (Emmaus, Pa.: Rodale Press, 1978).

7. *New Roots for Agriculture* (Andover, Mass: Brick House, 1980).

Bibliography

Allen, R. *How to Save the World: Strategy for World Conservation*. London: Kogan Page, 1980.

Altschul, A. M., and Wilcke, M. L. eds. *New Protein Foods*, vol. 3, *Animal Protein Supplies*. New York: Academic Press, 1978.

Anderson, W. *Diabetes: A Practical New Guide to Healthy Living*. New York: Arco, 1982.

Antibiotics in Animal Feed. Hearings Before the Subcommittee on Health and the Environment on H.R. 7285. Washington, D.C.: GPO, 1980.

Berry, W. *The Unsettling of America: Culture and Agriculture*. San Francisco: Sierra Club Books, 1977.

Bittinger, M. W., and Green, E. *You Never Miss the Water Till— (The Ogallala Story)*. Littleton, Colo.: Colorado Water Resources Publication, 1980.

Borgstrom, G. A. *Too Many: An Ecological Overview of Earth's Limitations*. New York: Collier, 1969.

Britton, D. K., and Hill, B. *Size and Efficiency in Farming*. Farnborough, England: Saxon House, 1975.

Caren, L. D. "Environmental Pollutants: Effects on the Immune System and Resistance to Infectious Disease." *Bioscience* 31 (1981): 582–86.

Carpenter, E. *Animals in Ethics*. London: Watkins, 1980.

Clark, S. R. I. *The Moral Status of Animals*. Oxford: Clarendon, 1977.

Comptroller General of the U.S. *Problems in Preventing the Marketing of Raw Meat and Poultry Containing Potentially Harmful Residues*. Washington, D.C.: General Accounting Office, 1979.

Council for Agricultural Science and Technology. *Scientific Aspects of the Welfare of Food Animals*. Report 91. Ames, Iowa: CAST, 1981.

Cunha, T. J. "The Animals as a Food Source for Man." *Feedstuffs*, May 31, 1982, 18–32.

Dahlberg, K.A., ed. *New Directions for Agriculture and Agricultural Research: Neglected Dimensions and Emerging Alternatives*. Totowa, N.J.: Rowman and Allanheld, 1985.

Davenport, C., et al. *The Effects of Tax Policy on American Agriculture*. USDA Agricultural Economic Report 480. Washington, D.C.: GPO, 1982.

Dawkins, M. S. *Animal Suffering: The Science of Animal Welfare*. London: Chapman and Hall, 1980.

Donham, K. J., Rubino, N., Thedell, T. D., and Kammermeyer, J. "Potential Health Hazard to Agricultural Workers in Swine Confinement Buildings." *Journal of Occupational Medicine* 19 (1977): 383–87.

Dunn, P. G. C. "Intensive Livestock Production: Costs Exceed Benefits." *Veterinary Record* 106 (1980).

Empty Breadbasket? The Coming Challenge to America's Food Supply and What We Can Do about It: A Study of the U.S. Food System. The Cornucopia Project of the Regenerative Agriculture Association. Emmaus, Pa.: Rodale Press, 1981.

Fox, M. W. *Factory Farming*. Washington, D.C.: The Humane Society of the United States, 1980.

———. *Farm Animals: Husbandry, Behavior and Veterinary Practice. Viewpoints of a Critic*. Baltimore: University Park Press, 1983.

———. *One Earth, One Mind*. New York: Coward, McCann and Geoghegan, 1980.

———. *Returning to Eden: Animal Rights and Human Responsibilities*. New York: Viking, 1980.

Gabel, M. *Ho-Ping: Food for Everyone*. New York: Anchor, 1979.

Godlovitch, S. R., and Harris, J., eds. *Animals, Men and Morals: An Enquiry into the Maltreatment of Non-humans*. New York: Taplinger, 1972.

Goldbeck, N., and Goldbeck, D. *The Supermarket Handbook: Access to Whole Foods*. New York: Harper & Row, 1974.

Goldschmidt, W. *As You Sow: Three Studies in the Social Consequences of Agribusiness*. New York: Allenheld, 1978.

Griffin, D. R. *The Question of Animal Awareness*. New York: Rockefeller University Press, 1981.

Hightower, J. *Eat Your Heart Out: How Food Profiteers Victimize the Consumer*. New York: Random House, 1976.

Hughes, K. *Return to the Jungle: How the Reagan Administration Is Imperiling the Nation's Meat and Poultry Inspection Program*. Washington, D.C.: Center for the Study of Responsive Law, 1983.

Hunter, B. T. *Consumer Beware! Your Food and What's Been Done to It*. New York: Simon & Schuster, 1971.

Jackson, W., Berry, W., and Colman, B. eds. *Meeting the Expectations of the Land: Essays in Sustainable Agriculture and Stewardship*. Berkeley, Calif.: North Point Press, 1985.

Knorr, D., ed. *Sustainable Food Systems*. Westport, Conn.: AVI Publishing, 1983.

Kramer, M. *Three Farms: Making Milk, Meat and Money from the American Soil*. Boston: Little, Brown, 1980.
Lappé, F. M. *Diet for a Small Planet*. New York: Ballantine, 1975.
______, and Collins, J. *Food First: Beyond the Myth of Scarcity*. Boston: Houghton Mifflin, 1977.
Lappé, M. *Germs that Won't Die*. New York: Anchor, 1982.
Linzey, W. *Animal Rights*. London: SCM Press, 1976.
Lockeretz, W., ed. *Environmentally Sound Agriculture*. New York: Praeger, 1983.
Madden, J. P. *Economics of Size in Farming*. USDA Agricultural Economic Report 107. Washington, D.C.: GPO, 1967.
Mason, J., and Singer, P. *Animal Factories*. New York: Crown, 1981.
Midgley, M. *Beast and Man: The Roots of Human Nature*. Ithaca, N.Y.: Cornell University Press, 1978.
Mooney, P. R. *Seeds of the Earth: A Private or Public Resource?* Ottawa: Inter Pares, Canadian Council for International Cooperation, 1980.
Morris, R. K., and Fox, M. W., eds. *On the Fifth Day: Animal Rights and Human Ethics*. Washington, D.C.: Acropolis, 1978.
Moss, R., ed. *The Laying Hen and Its Environment*. Boston: Martinus Nijhoff, 1980.
Murphy, L. B. *A Review of Animal Welfare and Intensive Animal Production*. Queensland, Australia: Queensland Department of Primary Industries, 1978.
National Research Council. *The Effects on Human Health of Subtherapeutic Use of Antimicrobials in Animal Feeds*. Washington, D.C.: National Academy of Sciences, 1980.
______, Committee on Diet, Nutrition, and Cancer. *Diet, Nutrition, and Cancer*. Washington, D.C.: National Academy of Sciences, 1982.
Null, G., and Null, S. *How to Get Rid of the Poisons in Your Body*. New York: Arco, 1977.
Office of Technology Assessment. *Drugs in Livestock Feed*, vol. 1, *Technical Report*. Washington, D.C.: GPO, 1979.
Paarlberg, D. *Farm and Food Policy: Issues of the 1980s*. Lincoln, Neb.: University of Nebraska Press, 1980.
Parham, B. *What's Wrong with Eating Meat?* Denver, Colo.: Ananda Marga Publications, 1981.
Paterson, D. A., and Ryder, R., eds. *Animal Rights: A Symposium*. London: Centaur, 1979.
Pimm, L. R. *Invisible Additives*. New York: Doubleday, 1981.
"Proper Diet Saves Lives, Land, Oil . . ." A symposium held by the American Association for the Advancement of Science at Toronto, Canada. Reported in *Science News* 119 (1981).

Public Voice for Food and Health Policy. *A Market Basket of Food Hazards: Critical Gaps in Government Protection*. Washington, D.C.: Public Voice for Food and Health Policy, 1983.

Purvis, M. J., and Otto, D. M. *Household Demand for Pet Food and the Ownership of Cats and Dogs: An Analysis of a Neglected Component of U.S. Food Use*. University of Minnesota Staff Paper Series, Department of Agricultural and Applied Economics. Minneapolis: University of Minnesota Press, 1976.

———. *Pet Food: How Much of a Drain on U.S. Food Supplies?* University of Minnesota Staff Paper Series, Department of Agricultural and Applied Economics. Minneapolis: University of Minnesota Press, 1978.

Ray, V. K. *The Corporate Invasion of American Agriculture*. Denver, Colo.: National Farmers' Union, 1968.

Regan, T. *All That Dwell Therein*. Berkeley: University of California Press, 1982.

———, and Singer, P., eds. *Animal Rights and Human Obligations*. Englewood Cliffs, N.J.: Prentice-Hall, 1976.

Regenstein, L. *America the Poisoned*. Washington, D.C.: Acropolis, 1982.

Rodale, R. *How Agriculture Hurts—And Can Help—The Soil*. Emmaus, Pa.: Rodale Press, 1981.

Rollin, B. *Animal Rights and Human Morality*. Buffalo, N.Y.: Prometheus, 1981.

Sainsbury, D. "The Influence of Environmental Factors on Livestock Health." In *Livestock Environment Affects Production and Health*. Proceedings of the International Livestock Environment Conference, American Society of Agricultural Engineers, held at the University of Nebraska–Lincoln, April 17–19, 1974, 4–13.

Schell, O. *Modern Meat: Antibiotics, Hormones and the Pharmaceutical Farm*. New York: Random House, 1984.

Schwabe, C. "The Holy Cow—Provider or Parasite? A Problem for Humanists." *Southern Humanities Review* 13 (1978): 251–78.

Siemens, L. B. "Ecological Agriculture." *World Agriculture* 29 (1980): 17–19.

Singer, P. *Animal Liberation*. New York: Random House, 1975.

Stone, C. D. *Should Trees Have Standing? Towards Legal Rights for Natural Objects*. Los Altos, Calif.: William Kaufmann, 1974.

Swanson, W., and Schultz, W. *Prime Rip: How the Meat Industry Cheats You Out of Up to Fifty Cents Per Pound on the Meat You Buy*. Englewood Cliffs, N.J.: Prentice-Hall, 1982.

Sybesma, W. *The Welfare of Pigs*. Boston: Martinus Nijhoff, 1981.

USDA, *Structure Issues of American Agriculture*. Economics, Statistics, and Cooperative Services Report 438. Washington, D.C.: GPO, 1979.

______. *A Time to Choose: Summary Report on the Structure of Agriculture*. Washington, D.C.: GPO, 1981.

______, Economics and Statistics Service. *Costs of Producing Livestock in the United States*. Prepared for the Committee on Agriculture, Nutrition, and Forestry, U.S. Senate. Washington, D.C.: GPO, 1981.

______, Team on Organic Farming. *Report and Recommendations on Organic Farming*. Washington, D.C.: GPO, 1980.

Who Will Control U.S. Agriculture? University of Illinois at Urbana-Champaign, College of Agriculture. Cooperative Extension Service Special Publication 28. 1980.

Who Will Sit Up with the Corporate Sow? Walthill, Neb.: Center for Rural Affairs, 1976.

Vallinatos, E. G. *Fear in the Countryside: The Control of Agricultural Resources in the Poor Countries by Non-Peasant Elites*. Cambridge, Mass.: Bullinger, 1976.

Supportive and Supplemental References

Consumer & Social Concerns

Adamson, R. H. (1991) Mutagens and carcinogens formed as a result of cooking meat. American Cancer Society. Phoenix, AZ.

Ahmed, A. K., Chasis, S., and B. McBarnette. (1984) *Petition of the Natural Resources Defense Council, Inc., to the Secretary of Health and Human Services Requesting Immediate Suspension of Approval of the Subtherapeutic Use of Penicillin and Tetracyclines in Animal Feeds.* New York: Natural Resources Defense Council.

Allen, L. H., et al. (1979) Protein-induced hypercaluria: a longer-term study. *The American Journal of Clinical Nutrition:* 32(741).

American Academy of Veterinary Preventive Medicine. (1985) *Continuing Education Seminar Proceedings: Meat/Poultry Microbiology: Disease Transmissions.* 29 October. Milwaukee, WI: American Academy of Veterinary Preventive Medicine.

American Cancer Society (pamphlet) (1987) *Cancer Facts and Figures 1987.* New York: American Cancer Society. p. 3.

Antibiotics in feed may help spread Salmonella, CDC says. (1989) *Feedstuffs.* 22 May.

Barkin, D., Batt, R. L., and B. R. DeWalt. (1990) *Food Crops vs. Feed Crops: Global Substitution of Grain in Production.* Boulder, CO: Lynne Riennen Publishers.

Beran, G. W. (1988) Use of drugs in animals, an epidemiological perspective. *Dollars and Sense: Proceedings of the Symposium on Animal Drug Use.* Rockville, MD: Center for Veterinary Medicine, Food and Drug Administration.

Brady, M. S., and S. E. Katz. (1988) Antibiotic/antimicrobial residues in milk. *Journal of Food Protection:* *51*(1). January.

Brody, J. E. (1990) Huge study of diet indicts fat and meat. *The New York Times*. 8 May. pp. 1, 14,15.

Cohen, M. L. (1986) Drug-resistant Salmonella in the United States: an epidemiological perspective. *Science: 234(2779)*. 12 November. pp. 964–9.

Collins-Thompson, D. L., Wood, D. S., and I. Q. Thomson. (1988) Detection of antibiotic residues in consumer milk supplies in North America using the charm test II procedure. *Journal of Food Protection: 51(8)*. August.

Committee on Diet, Nutrition, and Cancer: Assembly of Life Sciences, National Research Council, National Academy of Sciences. (1982) *Diet, Nutrition, and Cancer*. Washington, DC: National Academy Press.

Comptroller General of the United States. (1979) *Problems in Preventing the Marketing of Raw Meat and Poultry Containing Potentially Harmful Residues*. April 17. Washington, DC: General Accounting Office.

Concentration in meat packing. (1987) *CRA Newsletter*. August. Walthill, NE: Center for Rural Affairs.

Crawford, M. A. (1991) Fat animals—fat humans. *World Health*. July/August. pp. 23–5.

Donham, K. J. (1989) Relationships of air quality and productivity in intensive swine housing. *Agri-Practice: 10(6)*. November. pp. 15, 17.

Donham, K. J., et al. (1977) Potential health hazard to agricultural workers in swine confinement buildings. *Journal of Occupational Medicine: 19*. pp. 383–7.

El-Ahraf, A., et al. (1990) Dieldrin in the food chain. *Journal of Environmental Health*. July/August. pp. 17–9.

FDA Pesticide Program. *Residues in Foods 1990*. Washington, DC. pp. 1–21.

FDA studies sulfamethazine in milk. (1989) *FDA Veterinarian*. July/August. p. 3.

Feinman, S. E. (1984) The transmission of antibiotic-resistant bacteria to people and animals. In: (Steele, J. H. and G. W. Beran, [eds.]) *Zoonoses I*. CRC Handbook Series. Boca Raton, FL: CRC Press. pp. 151–71.

Flachowshy, G., and A. L. Hennig. (1990) Composition and digestibility of untreated and chemically treated animal excreta for ruminants—a review. *Biological Wastes: 31*. pp. 17–36.

Food and Drug Administration. (1989) Chlordane residues in broilers. *FDA Veterinarian*. March/April. p. 12.

Fox, M. W. (1990) *Inhumane Society: The American Way of Exploiting Animals*. New York: St. Martin's Press.

Fox, M. W. (1992) *Botswana's Cattle: Eden's End?* Washington, DC: The Humane Society of the United States.

Fromer, M. J. (1986) *Osteoporosis*. New York: Pocket Books. p. 9.

Gerber, D. B., et al. (1991) Ammonia, carbon monoxide, carbon dioxide, hydrogen sulfide, and methane in swine confinement facilities. *The Compendium: 13(9)*. September. pp. 1483–9.

Gimbutas, M. (1977) The first wave of Eurasian steppe pastoralists into copper age Europe. *Journal of Indo-European Studies 5*. Winter. pp. 277–338.

Goldschmidt, W. (1981) The failure of pastoral economic development programs in Africa. In: (Galaty, J. G., et al. ([eds.]) *The Future of Pastoral Peoples*. Proceedings of a conference held in Nairobi, Kenya, August 4–8. Ottawa, Canada: International Development Research Center.

Hindhede, M. (1920) The effect of food restriction during war on mortality in Copenhagen. *Journal of the American Medical Association: 74(6)*. pp. 381–2.

Hirsh, D. C., and N. Wigner. (1978) The effect of tetracycline upon the spread of bacterial resistance from calves to man. *Journal of Animal Science 46*. p. 1437.

Hoar, S. K., et al. (1986) Agricultural herbicide use and risk of lymphomas and soft-tissue sarcoma. *JAVMA: 256(9)*. pp. 1141–7.

Hoar, S. D., et al. (1988) A case-control study of non-Hodgkin's lymphoma and agricultural factors in eastern Nebraska. *Am. Journal of Epidemiology: 128(4)*. p. 901.

Holmberg, S. D., et al. (1984) Drug-resistant Salmonella from animals fed antimicrobials. *The New England Journal of Medicine: 311(10)*. pp. 617–22.

Howe, G., et al. (1991) A cohort study of fat intake and the risk of breast cancer. *Journal of the National Cancer Institute: 85*. pp. 5–8.

In-plant sulfa, antibiotic testing by USDA inadequate, OIG says. (1992) *Food Chemical News*. 20 January. pp. 46–7.

Institute of Medicine. (1989) *Human Health Risks with the Subtheraputic Use of Penicillin or Tetracyclines in Animal Feed*. Washington DC: National Academy Press. p. 2.

Johnson, N. E., et al. (1970) Effect of level of protein intake on urinary and fecal calcium and calcium retention of young adult males. *The Journal of Nutrition: 100(1425)*.

Jones, G. M., and E. H. Seymour. (1988) Cowside antibiotic residue testing. *Journal of Dairy Science: 71(6)*.

Junshi, C., et al. (1990) *Life-style, and Mortality in China: A Study of the Characteristics of 65 Chinese Counties*. New York: Oxford University Press.

Kaneene, J. B., and A. S. Ahl. (1987) Drug residues in dairy cattle industry: Epidemiological evaluation of factors influencing their occurrence. *Journal of Dairy Science: 70(10)*.

Kerr, L. A., et al. (1991) Aldicarb toxicosis in a dairy herd. *JAVMA: 198*. pp. 1636–9.

Levy, S. B. (1987) Antibiotic use for growth promotion in animals: ecologic and public health consequences. *Journal of Food Protection: 50(7)*. July. p. 616.

Listeria may be present in hot dogs. (1989) *JAVMA*. March 1. p. 626.

Lombardo, P. (1991) Pesticide residues in the U.S. diet. *Monitoring Dietary Intakes*. Springer-Verlag, NY: ILSI Monographs. pp. 183–90.

Lyons, R. W. (1980) An epidemic of resistant Salmonella in a nursery: animal-to-human spread. *Journal of the American Medical Association: 243(6)*. 8 February. pp. 546–7.

Mason, J. B. (undated) *Intensive Husbandry Systems, Animal Food Products and Human Health*. New York: The American Society for the Prevention of Cruelty to Animals.

Mathur, V. (1991) GAO says packing concentration not monitored. *Feedstuffs*. 4 November. p. 5.

McDonough, P. L., Jacobson, R. H., and J. F. Timoney. (1989) Virulence determinants of *Salmonella typhimurium* from animal sources. *American Journal of Veterinary Research: 50(5)*. May.

Miller, W. R. (1988) Violative drug residues. *Dollars and Sense: Proceedings of the Symposium on Animal Drug Use*. Rockville, MD: Center for Veterinary Medicine, Food and Drug Administration.

Mott L. (1984) *Pesticides in Food: What the Public Needs to Know*. San Francisco: Natural Resources Defense Council, Inc.

Murray, C. J. (1991) Salmonella in the environment. In: *Revue Scientifique et Technique: Animals. Pathogens and the Environment*. Paris, France: Office International des Epizootics. pp. 765–86.

Myers, C. F., Meek, J., Tuller, S., and A. Weinberg. (1985) Nonpoint sources of water pollution. *J. of Soil and Water Conservation: 40*. pp. 14–8.

Nair, P. P., et al. (1990) Influence of dietary fat on fecal mutagenicity in premenopausal women. *Int. Journal of Cancer: 46*. pp. 374–7.

National Cancer Institute. (1991) *Heterocyclic Aromatic Amines in Cooked Meats*. 25 March. Bethesda, MD: Office of Cancer Communications.

National Research Council. (1989) *Human Health Risks with the Subtherapeutic Use of Penicillin or Tetracycline in Animal Feed*. Washington, DC: National Academy Press.

National Research Council. Committee on Diet and Health, Food and Nutrition Board. (1989) *Diet and Health: Implications for Reducing Chronic Disease Risk*. Washington, DC: National Academy Press.

National Research Council. (1988) Chapter 2—Consumer concerns and animal product options. *Designing Foods: Animal Product Options in the Marketplace*. Washington, DC: National Academy Press. pp. 18–44.

National Research Council. (1987) *Poultry Inspection: The Basis for a Risk-Assessment Approach*. Washington, DC: National Academy Press. p. 6.

National Research Council. (1985) *Meat and Poultry Inspection*. Washington, DC: National Academy Press. p. 6.

National Research Council. (1980) *The Effects on Human Health of Subtherapeutic Use of Antimicrobials in Animal Feeds*. Washington, DC: National Academy Press.

Organization for Economic Cooperation and Development. (1986) *Water Pollution by Fertilizers and Pesticides*. Paris, France.

Oxby, C. (1989) *African Livestock-Keepers in Recurrent Crisis: Policy Issues Arising From the NGA Response*. London, England: International Institute for Environment and Development.

Phelps, A. (1990) Coccidiostats a health risk when broiler litter is fed. *Feedstuffs*. 17 September.

Pickrell, J. (1991) Hazards in confinement housing—gases and dusts in confined animal houses for swine, poultry, horses and humans. *Vet. Hum. Toxicology: 33(1)*. pp. 32–9.

Pimental, D. (1990) Environmental and social implications of waste in U.S. agriculture and food sectors. *J. Agric. Ethics: 3*. pp. 1–12.

Pimental, D. (1975) Energy and land constraints in food protein production. *Science: 190*. 27 November. pp. 754–61.

Pulce, C. (1991) Collective human food poisonings by clenbuterol residues in veal liver. *Vet. Hum. Toxicol: 33(5)*. pp. 480–1.

Radtke, T. M. Wastewater sludge disposal—antibiotic resistant bacteria may pose health hazard. *Journal of Environmental Health: 52(2)*. pp. 102–5.

Raghubir P., and S. and J. C. Street. (1980) Public Health Aspects of Toxic Metals in Animal Feeds. *JAVMA*. July 15. p. 149.

Reddy, B. S., et al. (1980) Nutrition and its relationship to cancer. *Advances in Cancer Research: 32(237)*.

Report of the Working Group on Arteriosclerosis of the National Institutes of Health. (1981) *Arteriosclerosis 1981*. Washington, DC: National Institutes of Health Publications. p. 523.

Riley, L. W., et al. (1983) Evaluation of isolated cases of salmonellosis by plasmid profile analysis: introduction and transmission of a bacterial clone by precooked roast beef. *The Journal of Infectious Diseases: 148(1)*. pp. 12–7.

Roberts, T. (1989) Human illness costs of foodborne bacteria. *American Journal of Agricultural Economics: 71(2)*. May. p. 471.

Roberts, T. L. (1980) Human illness costs of foodborne bacteria. *American Journal of Agr. Econ.: 71(2)*. pp. 468–74.

Scheid, J. F. (1991) AHI reports sales of U.S. animal drugs increased in 1990. *Feedstuffs*. 20 May. pp. 1, 23.

Schell, O. (1984) *Modern Meat: Antibiotics, Hormones, and the Pharmaceutical Farm*. New York: Random House.

Seymour, E. H., Jones, G. M., and M. L. Gilliard. (1988). Persistence of residues in milk following antibiotic treatment of dairy cattle. *Journal of Dairy Science: 71(8)*.

Sockett, P. N. (1991) A review: the economic implications of human Salmonella infection. *J. of Applied Bacteriology*. pp. 71, 289–95.

Spika, J. S., et al. (1987) Chloramphenicol-resistant Salmonella newport traced through hamburger to dairy farms. *The New England Journal of Medicine: 316(10)*. March 5. pp. 565–70.

Strauch, D. (1991) Survival of pathogenic micro-organisms and parasites in excreta, manure and sewage sludge. In: *Revue Scientifique et Technique: Animals, Pathogens and the Environment*. Paris, France: Office International des Epizootics. pp. 813–46.

Study reveals increasing rate of *Salmonella* excretion. (1989) *JAVMA: 195(1)*. 1 July.

Sun, M. (1984) Use of antibiotics in animal feed challenged. *Science: 226*. 12 October. pp. 144–6.

Tauxe, R. V. (1986) Antimicrobial resistance in human salmonellosis in the United States. *Journal of Animal Science: 62(Suppl.3)*. pp. 65–73.

Taylor, K. C. (1992) The control of bovine spongioform encephalopathy in Great Britain. *The Veterinary Record: 129*. pp. 522–6.

Troutt, F., et al. (1989) Antibiotics in beef production. *Chemical Use in Animal Production: Issues and Alternatives*. University of California, Agricultural Issues Center.

U.S. Congress, Office of Technology Assessment. (1988) *Pesticide Residue in Food: Technologies for Protection*. Washington, DC: U.S. Government Printing Office. October.

U.S. Department of Agriculture, Office of the Inspector General, Food Safety and Inspection Service. (1988) *Monitoring and Controlling Pesticide Residues in Domestic Meat and Poultry Products*. Atlanta, GA: U.S. Department of Agriculture, Office of the Inspector General. November. p. 9.

U.S. Environmental Protection Agency. (1984) *Report to Congress: Nonpoint Source Pollution in the U.S.* Washington, DC.

U.S. General Accounting Office (1992) *Food Safety and Quality: FDA Needs Stronger Controls Over the Approval Process for New Animal Drugs*. Washington, DC: U.S. General Accounting Office.

Van Dresser, W. R., and J. R. Wilcke. (1989) Drug residues in food animals. *JAVMA: 194(12)*. 15 June. p. 1701.

Voorburg, J. H. (1919) Pollution by animal production in the Netherlands: solutions. In: *Revue Scientifique et Technique: Animals, Pathogens and the Environment*. Paris, France: Office International des Epizootics. pp. 655–68.

Washington, G. E. (1988) Animal drug use considerations of the bovine practitioner. *Dollars and Sense: Proceedings of the Symposium on Animal Drug Use*. Rockville, MD: Center for Veterinary Medicine, Food and Drug Administration.

Willett, W. C., et al. (1990) Relation of meat, fat, and fiber intake to the risk of colon cancer in a prospective study among women. *The New England Journal of Medicine*. 13 December. pp. 1664–72.

World Health Organization (1990) *Public Health Impact of Pesticides Used in Agriculture*. Geneva: WHO.

Yuill, T. M. Animal diseases affecting human welfare in developing countries: impacts and control. *World Journal of Microbiology and Biotechnology*. pp. 157–63.

Environmental Issues

Benbrook C. (1991) *Sustainable Agriculture in the 21st Century: Will the Grass Be Greener?* Washington, DC: The Humane Society of the United States.

Blake, D. R. and F. S. Rowland. (1988) Continuing worldwide increase in tropospheric methane. *Science*. pp. 1129–31.

Bouwer, H. (1990) Agricultural chemicals and ground water quality—issues and challenges. *Ground Water Monitoring Review*. Winter.

Bower, H. and R. S. Bowman. (1989) *Agriculture Ecosystems and Environment: 26*. Amsterdam: Elsevier Science Publishers B. V. pp. 161–4.

Bowman, J. A. (1990) Ground-water-management-areas in United States. *Journal of Water Resources Planning and Management: 116(4)*. July/August. pp. 484–502.

Browder, J. O. (1988) Public policy and deforestation in the Brazilian Amazon. In: (Repetto, R. and Mr. Gillis, [eds.]) *Public Policies and the Misuse of Forest Resources*. New York: Cambridge University Press.

Brown, R. H. (1991) Environmental concerns make animal waste a target. *Feedstuffs*. 20 May. p. 9.

Buschbacher, R. J. (1986) Tropical deforestation and pasture development. *Bioscience: 36(1)*. January. pp. 22–8.

Byers, F. M., and N. D. Turner. (1991) The role of methane from beef cattle in global warming. *Beef Cattle Research in TX*. June. TX Agr. Exp. Stn. p. 69, PR-4838.

Caren, L. D. (1981) Environmental pollutants: effects on the immune system and resistance to disease. *Bioscience: 31*. pp. 582–6.

Clark, E. H. J. (1985) *Eroding Soils*. Washington, DC: The Conservation Foundation.

Daugherty, A. B. (1989) *U.S. Grazing Lands*. Statistical Bulletin No. 771. Washington, DC: U.S. Department of Agriculture.

Dregne, H. (1977) Desertification of arid lands. *Economic Geography*. October. pp. 322–31.

Environmental Protection Agency. (1990) *Policy Options for Stabilizing Global Climate*. Washington, DC: U.S. Environmental Protection Agency.

Follett, R. F. (ed.) (1989) Nitrogen management and ground water pollution. *Developments in Agriculture and Managed-Forest Ecology: 21*. pp. 35–74.

Fox, M. W. (1992) *Botswana's Cattle: Eden's End?* Washington, DC: The Humane Society of the United States.

Gilbertson, G. B., et al. (1981) *Controlling Runoff for Livestock Feedlots*. October. Agricultural Research Service, Bulletin No. 441. Washington, DC: U.S. Department of Agriculture.

Gilliom, R. J. and D. Clifton (1990) Organochlorine pesticide residues in bed sediments of the San Joaquin River, CA. *Water Res. Bulletin: 26(1)*. February. pp. 11–23.

Grandin, B. E. (1986) Human demography and culture: factors in range management. *Wildlife/Livestock Interfaces on Rangelands*. Nairobi, Kenya: Inter-African Bureau of Animal Resources, PO Box 30786. pp. 119–27.

Halstead, J. M., et al. (1990) Ground water contamination from agricultural sources: implications for voluntary policy adherence from Iowa and Virginia farmers' attitudes. *America Journal of Alt. Agriculture*. 30 May. pp. 126–33.

Hecht, S. B. (1989) The sacred cow in the green hell: livestock and forest conversion in the Brazilian Amazon. *The Ecologist: 19*. pp. 229–34.

Hodges, R. D., and A. M. Scofield. (1983) Agricologenic disease—a review of the negative aspects of agricultural systems. *Biological Agriculture and Horticulture: 1*. pp. 269–325.

Hubert, C. (1991) Spring rains found tainted by herbicides. *The Des Moines Register*. 27 April. p. 1.

Jones, J. R. (1990) *Colonization and Environment: Land Settlement Projects in Central America*. Tokyo: U.N. University Press.

Kerr, L. A., et al. (1991) Chronic copper poisoning in sheep grazing pastures fertilized with swine manure. *JAVMA: 198(1)*. pp. 99–101.

Lashof, D. A., and D. A. Tirpak (eds.) *Policy Options for Stabilizing Global Climate*. February. Washington, DC: U.S. EPA Office Policy, Planning and Evaluation. p. 55.

McKinney, T. R., and A. Gold. (1987) Effect of water pollution control on concentration. *United States Feedlots*. November 5. Snowmass, CO: Rocky Mountain Institute.

McNaughton, S. (1990) Mineral nutrition and seasonal movements of African migratory ungulates. *Nature: 345*. pp. 613–5.

Midwest Plan Service Committee. (1985) *Livestock Waste Facilities Handbook*. Ames: Iowa State University.

Mosler, A.D., Schimel, D., Valentine, D., and K. Bronson. (1991) Methane and nitrous oxide fluxes in native, fertilized and cultivated grasslands. *Nature 350*. pp. 330–2.

Muirhead, S. (1990) Nutrition and health: strategies available for reducing ruminant methane emissions. *Feedstuffs*. November 12. pp. 10, 22.

Myers, C. F., Meek, J., Tuller, S., and A. Weinberg. (1985) Nonpoint sources of water pollution. *J. of Soil and Water Conservation: 40*. pp. 14–8.

National Research Council. (1986) *Pesticides and Groundwater Quality: Issues and Problems in Four States*. Washington, DC: National Academy Press.

Nations, J. D. and D. I. Komer. (1983) Central America's tropical rainforests: positive steps for survival. *Ambio: 12(5)*.

Pearce, F. (1986) Are cows killing Britain's trees? *New Scientist*. 23 October. p. 20.

Phelps, A. (1989) Dutch government puts ceiling on swine production to prevent pollution problems. *Feedstuffs*. 23 October. p. 5.

Picard, L. A. (1987) *The Politics of Development in Botswana: A Model for Success?* Boulder, CO: Lynne Reinner Publishers.

Pimental, D. (1990) Environmental and social implications of waste in U.S. agriculture and food sectors. *J. Agric. Ethics: 3.* pp. 1–12.

Pimental, D., et al. (1980) Environmental and social costs of pesticides: a preliminary assessment. *Oikos 34.* pp. 127–40.

Poffenberger, M. (1990) *Joint Management for Forest Lands: Experiences from South Asia.* New Delhi: Ford Foundation.

Power, J. F., and J. S. Schepers. (1989) Nitrate contamination of groundwater in North America. *Agriculture, Ecosystems and Environment: 26.* pp. 165–87.

Radtke, T. M. (1989) Wastewater sludge disposal—antibiotic resistant bacteria may pose health hazard. *Journal of Environmental Health: 52(2).* pp. 102–5.

Sandford, S. (1983) *Management of Pastoral Development in the Third World.* Chichester, England: John Wiley and Sons.

Saull, M. L. (1990) Nitrates in soil and water. *New Scientist—Inside Science.* 15 September. pp. 1–4.

Saunders, D. A., et al. (1990) *Australian Ecosystems: 200 Years of Utilisation, Degradation and Reconstruction.* Proceedings of the Ecological Society of Australia, Ecological Society of Australia, Chipping Norton, Australia.

Strauch, D. (1991) Survival of pathogenic micro-organisms and parasites in excreta, manure and sewage sludge. In: *Revue Scientifique et Technique: Animals, Pathogens and the Environment.* Paris, France: Office International des Epizootics. pp. 813–46.

U.S. Environmental Protection Agency. (1984) *Report to Congress: Nonpoint Source Pollution in the U.S.* Washington, DC.

U.S. Forest Service, USDA. (1988) *An Assessment of the Forest and Range Land Situation in the United States.* Washington, DC: Government Printing Office.

U.S. General Accounting Office. (1991) *Rangeland Management BLM's Hot Desert Grazing Program Merits Reconsideration.* Washington, DC: U.S. General Accounting Office.

Voorburg, J. H. (1991) Pollution by animal production in The Netherlands. In: *Revue Scientifique et Technique: Animals, Pathogens and the Environment.* Paris, France: Office International des Epizootics. pp. 655–68.

Wald, J., and D. Alberswerth. (1989) *Our Ailing Public Range Lands.* Washington, DC: National Wildlife Federation and Natural Resources Defense Council.

Williamson, D., and J. Williamson. (1984) Botswana's fences and the depletion of Kalahari wildlife. *Oryx: 18.* pp. 218–22.

Farm Animal Health & Welfare

Baker, F. H., and N. Ramm. (1989) The role and contributions of animals in alternative agricultural systems. *American Journal of Alternative Agriculture 4: (3/4).*

Barbano, D. M., et al. (1987) Impact of mastitis on dairy product yield and quality. In: *Proceedings of the 26th Annual Meeting of the National Mastitis Council.* Arlington, VA: National Mastitis Council.

Burris, R. (1989) Cited in: One million calves die annually from weaning stress. *Cattle Today*. December 16. p. 6.

Distl, O., et al. (1989). Analysis of relationships between veterinary recorded production diseases and milk production in dairy cows. *Livestock Production Science: 23*. pp. 67–78.

Flachowshy, G., and A. L. Hennig. (1990) Composition and digestibility of untreated and chemically treated animal excrete for ruminants—a review. *Biological Wastes: 31*. pp. 17–36.

Friend, T. Dellmeier, G. R., and E. E. Gbur. (1985) *Comparison of Four Methods of Calf Confinement. 1. Physiology. Technical Article 18960.* College Station, TX: Texas Agricultural Experiment Station.

Grandin, T. (1990) Calf-handling needs improvement. *Meat & Poultry*. July. p. 88.

Halverson, M. (1991) *Farm Animal Welfare: Crisis or Opportunity for Agriculture?* Staff Papers Series. St. Paul: Department of Agricultural and Applied Economics, University of Minnesota, Institute of Agriculture, Forestry and Home Economics. Staff Paper P91-1.

Hill, M. A. (1990) Economic relevance, diagnosis, and countermeasures for degenerative joint disease (osteoarthrosis) and dyschondroplasia (osteochondrosis) in pigs. *JAVMA: 197(2)*. pp. 245–9.

Julian, R. J. (1990) Pulmonary hypertension: A cause of right heart failure, ascites in meat-type chickens. *Feedstuffs*. Jan. 29. pp. 19, 21–22, 78.

Kerr, L. A., et al. (1991) Chronic copper poisoning in sheep grazing pastures fertilized with swine manure. *JAVMA: 198(1)*. pp. 99–101.

Killingsworth, M. L., and J. B. Kliebenstein. (1984) Estimation of production cost relationships for swine producers using differing levels of confinement. *J. of the Am. Soc. of Farm Managers and Rural Appraisers: 48(2)*. pp. 32–6.

Kliebenstein, J. B., Kirtley, C. L., and M. L. Killingsworth. (1981) A comparison of swine production costs for pasture, individual, and confinement farrow-to-finish production facilities. *Special Report 273*. Columbia, MO: Agricultural Experiment Station, University of Missouri-Columbia.

Largest slaughter check ever finds respiratory disease to be widespread. (1989) *Feedstuffs*. 7 January. p. 8.

Lidvall, E. R., Ray, R. M., Dixon, M. C., and R. L. Wyatt (1980) *A Comparison of Three Farrow-Finish Pork Production Systems.* Reprint from Tennessee Farm and Home Science No. 116.

Mason, J. B. (undated) *Intensive Husbandry Systems, Animal Food Products and Human Health.* New York: The American Society for the Prevention of Cruelty to Animals.

Mickley, L. D., and M. W. Fox. (1987) The case against intensive farming of food animals. In: (Fox, M. W. and L. D. Mickley. [eds.]) *Advances in Animal Welfare Science 1986/1987*. Boston: Martinus Nijhoff.

Muntz, S. H. (1991) Range layers compared to confined layers at rock creek farm. *Missouri Farm*. May/June. p. 24

National Mastitis Council. (1987) *Current Concepts of Bovine Mastitis. 3rd ed.* Arlington, VA.

Norgaard-Nielsen, G. (1990) Bone strength of laying hens kept in an alternative system compared with hens in cages and on deep-litter. *Britain Poultry Science: 31*. pp. 81–9.

Paape, M. J. (1986) Inducing natural defense mechanisms to promote mastitis control. *Proceedings of the 18th Annual Convention of the American Association of Bovine Practitioners*. April.

Phelps, A. (1990) Coccidiostats a health risk when broiler litter is fed. *Feedstuffs*. 17 September.

Pickrell, J. (1991) Hazards in confinement housing—gases and dusts in confined animal houses for swine, poultry, horses and humans. *Vet. Hum. Toxicology: 33(1)*. pp. 32–9.

Pursel, V. G., et al. (1989) Genetic engineering of livestock. *Science*. 16 June. p. 1285.

Reid, I. M., et al. (1983) Immune competence of dairy cows with fatty liver. In: Swedish University of Agricultural Sciences, *Proceedings 5th Int. Conf. on Production Diseases in Farm Animals.*

Royal Society for Prevention of Cruelty to Animals (RSPCA). (1989) *Osteopenia in Laying Hens*. Report on a conference sponsored by the RSPCA at Streatley on Thames, April 4th. London, England.

Salman, M. D., et al. (1991) Annual costs associated with disease incidence and prevention in Colorado cow-calf herds participating in rounds 2 and 3 of the NAHMS from 1986 to 1988. *JAVMA: 198(6)*. pp. 968–73.

Study reveals increasing rate of *Salmonella* excretion. (1989) *JAVMA: 195(1)*. 1 July.

United States Department of Agriculture: APHIS: (1991) *National Swine Survey: Morbidity/Mortality and Health Management of Swine in the United States*. November. Fort Collins, CO. pp. 1, 11–14.

Livestock & Sustainability Issues

Baker, F. H., and R. J. Jones. (1985) *Proceedings on Multi-species grazing*. Morrilton, AR: Winrock International.

Beauchamp, E. G. (1990) Animals and soil sustainability. *Journal of Agricultural Ethics: 3*. pp. 89–98.

Benbrook, C. (1991) *Sustainable Agriculture in the 21st Century: Will the Grass Be Greener?* Washington, DC: The Humane Society of the United States.

Blaser, R. E., et al. (1980) Forage-animal systems for economic calf production. In: *Proceedings of the XIII International Grassland Congress*. Berlin: Akademi-Verlag. pp. 667–71.

Bochncke, E. (1985) The role of animals in biological farming system. In: (Eden, T., et al. [eds.]) *Sustainable Agriculture and Integrated Farming Systems*. Michigan State University Press.

Chaney, E., et al. (1990) *Livestock Grazing on Western Riparian Areas*. Eagle, ID: Northwest Resource Information Center.

Cramer, C. (1991) Pastures beat BGH. *The New Farm*. July/August. pp. 18–22.

de Haan, C. (1990) Changing trends in the world bank's lending program for rangeland development. *Low Input Sustainable Yield Systems: Implications for the World's Rangelands*. Proceedings of the 1990 International Rangeland Development Symposium, Reno, NV. February 15. Logan, UT: Utah State University, Department of Range Science.

De Vries, J. (1990) Zero grazing: successfully using livestock in regenerative farming system. *VITA News* (Volunteers in Technical Assistance). April. Arlington, VA.

Doescher, P. S., Tesch, S. D., and M. A. Castro. (1987) Livestock grazing: a silvicultural tool for plantation establishment. *Journal of Forestry: 10*. p. 2,937.

Doran, J. W., et al. (1987) Influence of alternative and conventional agricultural management on soil microbial processes and nitrogen availability. *Am. J. of Alternative Agric.: 2(3)*. pp. 99–106.

Dunn, P. G. C. (1980) Intensive livestock production: costs exceed benefits. *Veterinary Record: 106*.

Glenn, J. C. (1988) Discussion paper 39. *Livestock Production in North Africa and the Middle East: Problems and Perspectives*. Washington, DC: World Bank.

Goldschmidt, W. (1981) The failure of pastoral economic development programs in Africa. In: (Galaty, J. G., et al., [eds.]) *The Future of Pastoral Peoples*. Proceedings of a conference held in Nairobi, Kenya, August 4–8. Ottawa, Canada: International Development Research Center.

Graham, O. (1989) A land divided: the impact of ranching on pastoral society. *The Ecologist*. September/October.

Grandin, B. E. (1986) Human demography and culture: factors in range management. *Wildlife/Livestock Interfaces on Rangelands*. Nairobi, Kenya: Inter-African Bureau of Animal Resources, P.O. Box 30786. pp. 119–27.

Groh, T. M., and S. H. McFadden. (1990) *Farms of Tommorow*. Kimberton, PA: Bio-Dynamic Farming and Gardening Association, Inc.

Gussow, J. D. (1991) *Chicken Little. Tomato Sauce and Agriculture*. New York: The Bootstrap Press.

Heady, E. O., and H. R. Jensen. (1951) The economics of crop rotations and land use: a fundamental study in efficiency with emphasis on economic balance of forage and grain crops. *Research Bulletin 383*. August. Ames, IA: Agricultural Experiment Station, Iowa State University.

Hodges, R. D., and A. M. Scofield. (1983) Effect of agricultural practice on the health of plants and animals produced: A review. In: (Lockeretz, [ed.]) *Environmentally Sound Agriculture: Selected Papers from the Fourth International Conference of the International Federation of Organic Movements.* New York: Praeger.

Honeyman, M. (1991) Sustainable swine production in the U.S. corn belt. *American Journal of Alternative Agriculture: 6(2).* pp. 63–9.

Karstad, L. (1986) Can livestock and wildlife co-exist? *Wildlife/Livestock Interfaces on Rangelands.* Nairobi, Kenya: Inter-African Bureau of Animal Resources, P.O. Box 30786. pp. 51–5.

Kliebenstein, J. B., Kirtley, C. L., and M. L. Killingsworth. (1981) A comparison of swine production costs for pasture, individual, and confinement farrow-to-finish production facilities. *Special Report 273*. Columbia, MO: Agricultural Experiment Station, University of Missouri.

Lang, R. A. (1990) *Improving Ruminant Production and Reducing Methane Emissions from Ruminants by Strategic Supplementation* (draft). Armisdale, Australia: Institute of Biotechnology, University of New England.

Livingston, I. (1991) Livestock management and "overgrazing" among pastoralists. *Ambio: 20.* pp. 80–5.

Mosler, A.D., et al. (1991) Methane and nitrous oxide fluxes in native, fertilized and cultivated grasslands. *Nature: 350.* pp. 330–2.

Oxby, C. (1989) *African Livestock-Keepers in Recurrent Crisis: Policy Issues Arising from the NGA Response.* London, England: International Institute for Environment and Development.

Pearson, H. A., et al (eds). (1991) *Development or Destruction? The Conversion of Forest to Pasture in Latin America.* Boulder, CO: Westview Press.

Preston, T. R. (undated) Strategies for livestock production in the tropical third world. *Livestock Production.* Cali, Columbia: CIPAV. p. 208.

Preston, T. R., and I. Vacarro. (1989) Dual purpose cattle production systems. In: (Phillips, C. J. C. [ed.]). *New Techniques in Cattle Production.* London: Butterworths Scientific. pp. 20–32.

Reagonold, J. P., Elliott, L. F., and Y. L. Unger. (1987) Long-term effects of organic and conventional farming on soil erosion. *Nature: 330.* pp. 370–2.

Reisner, M., and S. Bates. (1990) *Overtapped Oasis: Reform or Revolution for Western Water.* Washington, DC: Island Press.

Royal Agricultural Society of England. (1991) *The State of Agriculture in the United Kingdom.* Stoneleigh Park, Warwickshire: National Agricultural Center.

Russell, J. R., and M. R. Brasche. (1990) Letting cows harvest forage year-round. *Leopold Letter: A Newsletter of Leopold Center for Sustainable Agriculture: 2(3)*. Ames, IA: Leopold Center for Sustainable Agriculture, Summer. pp. 4–7.

Sandford, S. (1983) *Management of Pastoral Development in the Third World*. Chichester, England: John Wiley and Sons.

Shrader, W. D. and R. D. Voss. (1980) Soil fertility: crop rotation vs. monoculture. *Crops and Soils Magazine:* 7. pp. 15–8.

Smith, S. R., Jr., Bouton, J. H., and C. S. Hoveland. (1989) Alfalfa persistence and regrowth potential under continuous grazing. *Agron. J.: 81*. pp. 960–5.

Sweeten, J. M. (1990) Water use, animal waste, and water pollution. In: (Cross, H. R., and F. M. Byers, [eds.]). *Current Issues in Food Production: A Perspective on Beef as a Component in Diets for Americans*. Englewood, CO: National Cattlemen's Association.

Talbot, L. M., Payne, W. J. A., Ledger, H. P., Verdcourt, D., and M. H. Talbot. (1965) The meat production potential of wild animals in Africa: a review of biological knowledge. *Tech. Commun. Commonwealth Bur. Anim. Breed. Genet: 16*. pp. 1–42. Furnham Royal: Commonwealth Agricultural Bureau x.

Topel, D. G. (1986) The beef industry and its future. *Forages: The Grassroots of Agriculture*. Proceedings, 1986 Forage and Grassland Conference, Athens, GA. American Forage and Grassland Council and the University of Georgia.

United Nations (UN) Food and Agriculture Organization (FAO). (1990) *Production Yearbook 1989*. Rome, Italy.

Vera, R. R., et al. (1984) Development of improved grazing systems in the savannas of tropical America. In: *Rangelands: A Resource Under Siege*. Proceedings of the 2nd International Rangeland Congress, Adelaide, Australia. New York: Cambridge University Press.

Ward, G. M., et al. (1977) Beef production options and requirements for fossil fuel. *Science: 198*. pp. 265–71.

Whose Common Future: Northern Agriculture. (in press) *The Ecologist*.

Wolfshohl, K. (1991) Pink veal finds a growing market. *Progressive Farmer*. August. pp. 34–5.

Young, D. L., and K. M. Painter. (1990) Farm program impacts on incentives for green manure rotations. *American Journal of Alternative Agriculture: 5(3)*. pp. 99–105.

Younos, T. M. (1990) Integrated manure management. *Agricultural Engineering*. February. St. Joseph, MI.

Index

Acid rain, 58
Aflatoxins, 72, 77, 96, 107
Agribiotechnology, 185–195
Agribusiness, definition of, x, 24
Agricide, definition of, 50
Agriculture, 128; efficiency, 62–63; reforms, 153–162; 163–217
Agripower, 31
Air pollution, 58
Albedo, 63
Allergies: to antibiotics, 90; to food, 105, 189
American Farm Bureau Federation, 33–34, 54–55, 57, 104
Animal health costs, 36–37, 46, 212
Animal husbandry, 1–25; humane systems, 8, 123–126
Animal liberation, 109–110
Animal rights, 109–114, 117–118
Animal welfare and economics, 35–37
Animal welfare codes, 155
Anthropocentrism, 111, 114
Antibiotic resistance, 206
Antibiotics: in feed, 12–19, 66–67, 79, 139–140; resistance to, in bacteria, 88–94, 206
Arteriosclerois, and diet, 103, 104, 105

Banking policies, 34
Beef cattle welfare, 7–8, 119, 125
Biodiversity, 170–171, 179, 182
Bioethical criteria, 193–194
Biotechnology, 171–172, 185–195
Blood pressure, and diet, 103
Bovine growth hormone (rBGH), 190, 205
Boycott, of hamburgers, 52–53
Breast cancer, and diet, 103
Breast milk, contamination of, 69–70
Byproducts, in feed, 20–21, 65, 77, 88; pet foods, 129–135

Cadmium, toxicity, 75–76
Calcium, in diet, 104–105
Cancer: chicken meat, 95; and diet, 203–204; environmental factors, 67–68, 72, 75; in farmers, 40. *See also* Pesticides
Cartesian attitude, 112
Cattle: environmental impact of, 53–54, 56; range-raising of, 55–56
Chickens, welfare of, 4–5, 119, 123, 125
Cholesterol, 203
Condemnation, of meat, 10

Confinement systems, economics, 137–138
Conservation tillage, 51
Controlled grazing, of cattle, 55–56
Corporate control of agriculture, 22–24, 27
Corporate monopoly, 151
Corporate peonage, 34, 38–39
Council for Agricultural Science and Technology (CAST), 32–33
Cows (dairy), welfare of, 119, 123, 126
Creationism, 113
Crop losses, due to pests, 140

Dairy support payment, 44–45
Deforestation, 52–53, 64
DES, 77–78
Diabetes, 202
Diet: ancestral, 106–107; of farm animals, 11, 19–20, 21; and health, 100–107, 202–206; humane, 120–129, 201, 213–214
Dioxins, 68–69
Disease and husbandry, 167
Diversity, preservation, 170
Dominion, 113
Drugs: in "factory" farming, 10, 18–19; justifications for use in feed, 27; residues, 11, 66–67, 76, 80–82; sales, 18–19

Eating with Conscience, 201, 213
Ecology, of farming, 121
Economics, of farming, 26–49, 136–139, 145–147
Efficiency of farming systems, 136–139, 145–147
Eggs: "factories," 4; quality of, 101–102
Energy consumption, 141–143; in food production, 62–63, 207–208
Entropy, 50, 121
Eutrophication, 148
Exports: limits of, 153; of produce, 31–32

"Factory" farming, 1–25, 118–120; costs of, 138
Famine, 163–183
Farm animals: numbers raised, 22; welfare of, 108–129, 144–146, 211–212
Farmer's Home Administration, 29–30, 46
Farming: efficiency, 136–139, 145–147; reforms, 153–162, 186–201
Farms: categories of, 21–22; growth of, 42–43; optimal size, 137
"Fast-food" industry, 100
Fat and health, 103–104, 106–107, 203
Federal inspection costs, 45
Federal support programs, 27–30
Feed additives. *See* Antibiotics; Drugs
Fertilizers, 20; and crop diseases, 100–101; health risks of, 74–76; and productivity, 140; synthetic, 59, 74–75
Fish, 127–129
Fluoride, toxicity of, 76–77
Food: additives, 66; allergies, 105; chemical residues in, 11, 64–73, 74, 75–76, 80–82; irradiation, 95–98; labeling, 189, 195; nutritional quality of, 100; for pets, 129–135; poisoning, 79, 82–88, 204–205; prices, 43; production, 99. *See also* Pesticides
Fossil fuel, consumption of, 62–63, 141–144, 207

Generic pet foods, 132
Genetic engineering, 185–195
Grain use, 207
Grazing, 209
"Greenhouse" effect, 63
"Green Revolution," 149, 163–164
Groundwater supplies, 61. *See also* Water
Growth hormones, 77–78

Health and "Connectedness," 200–201
Health problems, of farmers, 40–41
Heart disease, 204
"Hog" hotels, 24–25
Holistic range management, 211
Homogenized milk, 105
Honey bees, importance of, 57
Humane alternatives, 8, 125–126
Humane sustainable agriculture, 183–184, 206, 210–211, 213
Hydroponics, 158–159

Immunosuppression, 69–70
Integrated pest management, 156
Interest rates, 27
Irradiation of food, 96–98, 206
Irrigation, 61; subsidies, 48

Kosher slaughter, 118

Lactose intolerance, 105
Land-leasing, 42
Land loss, 173
Land use, 51–52
Laying hens, welfare of, 4, 119, 123, 126
Lead, in bone meal, 104
Leukemia, 95
Livestock: condemnation of, 87; crop integration, 168–169, 175; grazing, 169; losses of, 37, 46; in third world, 166–167, 175–179

Manure: disposal of, 57–58; pollution, 148, 208
Marketing cooperatives, 157
Marketing order, 27–28
Meat consumption, and health, 102–103
Meat inspection, inadequacies of, 41, 81–88, 94
Meat production, reforms, 153
Medicated feeds. *See* Antibiotics; Drugs
Migrant workers, 48–49
Milk, health hazards of, 104–105. *See also* Leukemia
Mortality rates, of farm animals, 9–10

Native peoples, 163–181
Nitrofurans, 79
Nitrosamines, 74–75
Nutrient deficiencies, 60
Nutrient value of foods, 100–102
Nutritional quality, of hydroponic crops, 159

Ogallala aquifer, 61, 63
Organic agriculture, and biotechnology, 191–193
Organic farming, 158–161

Pastoralists, 178–179
Patents, 171, 186
Payment in kind (PIK) programs, 30
Pesticides, 20, 38, 57, 65–73, 140–141, 174; insect resistance to, 73–74; and pollution, 73
Pests, and crop losses, 140

Pet foods, 129–135
Petrochemical industry, 40
Phosphates, in food, 76
Pigs, welfare of, 5–6, 118–119, 123, 126
Plastic, health hazard, 64
Pollution: agrichemical, 57; air, 58, 73; manure, 57–58, 61; rain, 58; selenium, 59; water, 61, 68, 72, 74–75, 208
Poultry, welfare of, 4–5, 119, 123, 125
Poverty, 163–183
Predator control, 55, 180
Production costs, 136–139, 145–147
Protein consumption, 99, 203

Reforms, in livestock production, 153, 183–184
Regenerative agriculture, 121
Reye's syndrome, 70
Rotational grazing, 211

Salmonella, food poisoning, 7, 82–88; outbreaks of, 92–94
Seed stock, monopoly of, 22–23
Selenium, toxicity of, 59
Sheep, welfare of, 119, 123, 125
Slaughter, humane, 121
Slaughterhouse inspection, 41, 81–88, 94
"Sodbusters," 51
Soil erosion, 51, 60–61, 173–174
Sows, welfare of, 6, 118–119, 126

Tax policies, 28–30
Tax shelters, 29, 46
Thelarche, 78
Third world, 150–151
Trace minerals: deficiencies of, 59–61; depletion of, 58, 63; toxicity of, 76
Transgenic animals, 192–193
Transgenic crops, 193
Transportation: of animals, 37, 121; costs of, 46; livestock stress and, 86
Trophobiosis, 100
Tropical rain forest, destruction of, 52–53

Urbanization, 175–176

Veal calves, welfare of, 6–7, 119, 123, 125
Vegetarianism, 109, 112, 115, 214–217; cats and dogs and, 134–135; and health, 102–104; and water conservation, 61
Vertical integration, 23
Virus infections, 94–95

Water: consumption of, 61–62, 207; irrigation, 61; pollution, 61, 68–72, 74–75
Weather, control of, 62
Welfare, of farm animals, 108–129
Wild game, 129
Wildlife: destruction of habitat, 52, 54, 56, 112; displacement of, 181, 215; ranching of, 53–54, 150, 178
World hunger, 163–183
World population, 173

Yeast infections, 90–91, 106